The Cosmogony of the Solar System

The Cosmogony of the Solar System

by
Fred Hoyle

Honorary Professorial Fellow in the
Department of Applied Mathematics
at University College, Cardiff, Wales

Enslow Publishers
Bloy Street and Ramsey Avenue
Hillside, New Jersey 07205

First American publication, 1979

© 1978 Fred Hoyle

First published in U.K. in 1978 by University College Cardiff Press

Library of Congress Cataloging in Publication Data
Hoyle, Fred, Sir.
   The cosmogony of the solar system.

   Includes index.
   1.   Solar system - Origin.   1.   Title.

QB501.H79   521'.54   78-21286
ISBN 0-89490-023-4

Printed in the U. S. A.

TO MARIA AND ERIC MUHLMANN

Other titles by Sir Fred Hoyle

1949	Some Recent Researches in Solar Physics—Cambridge University Press
1950 (revised 1960)	The Nature of the Universe—Heinemann & Harpers (Blackwells & Harpers)
1953	A Decade of Decision—Heinemann
1955	Frontiers of Astronomy—Heinemann & Harpers
1956	Man and Materialism—Harpers
1962	Astronomy—Rathbone Books
1965	Galaxies, Nuclei and Quasars—Heinemann & Harpers
1966	Man in the Universe—Columbia University
1966	Galaxies, Nuclei and Quasars—Heinemann & Harpers
1972	From Stonehenge to Modern Cosmology—W. H. Freeman
1973	Nicolaus Copernicus—Heinemann & Harpers
1974	Action at a Distance in Physics and Cosmology (with J. V. Narlikar)—W. H. Freeman
1974	Astronomy and Cosmology—W. H. Freeman
1975	Highlights in Astronomy—W. H. Freeman & Heinemann
1976	On Stonehenge—W. H. Freeman & Heinemann
1976	Ten Faces of the Universe—W. H. Freeman & Heinemann
1977	Energy or Extinction—Heinemann

Contents

Contents

Preface

PREFACE

No attempt has been made in this book to include any of the exquisite detail of modern isotope chemistry, which is sometimes thought to be essential to an understanding of the origin of the solar system. My reason is that other facts concerning the structure and dynamics of the planets of a more sledgehammer quality have seemed to me to be too often ignored. These other facts alone appear to justify a short book in their own right.

I have unashamedly written in old-style c.g.s. units. For those unfortunate young people 'brought up' on SI units, and who are thence largely disbarred from understanding quantum theory, relativity, and modern high-energy physics (as well as being cut-off from the main stream of scientific literature) it will probably be sufficient to know that 10^2 cm $= 1$ m, 10^3 g $= 1$ kg, 1 Å $= 10$ nm,

Fred Hoyle

Dockray, Cumbria

Introduction

The French mathematician Pierre-Simon Laplace (1749–1827) lived in the so-called age of reason, when it was believed that subject to a few small addenda (like magnetism and voltaic electricity) the laws governing the universe were well-known. We suffer from the same illusion today. So, thought Laplace, everything that happens is calculable, from the motions of the stars in the heavens down to the twinkle in my lady's eye. Laplace conceived of a supermathematician who proceeded to work out all the future, and for whom life would no doubt be very dull, once the first flush of success at the horse races had worn off.

In physics today we distinguish between problems involving just a few particles and problems involving many particles. The first of these categories is relatively simple and so can be studied in depth. From its study we believe the laws of the universe can be discovered in their entirety. And then, from the laws so discovered, problems in the second category can be calculated in the manner of Laplace's supermathematician. Just from the laws of the basic particles, all of chemistry, biochemistry, biology, geophysics, and astronomy can be worked out. Needless to say, such a programme has not yet been carried through, and there are reasonable grounds for doubting that it will ever be carried through. I do not mean to imply that something new and subtle inserts itself (as quite a few scientists seem to believe) between problems in the small and those in the large. My objection to

1

supercalculation is entirely mundane. It is simply that there is no practical way to do it. Given the most efficient practical calculator, my hunch is that to calculate everything which has happened on the Earth throughout its history would take longer than the age of the Earth, and to store away the results of such a calculation would occupy more space than the bulk of the whole Earth. Correction! Given the most efficient calculator it would take precisely the age of the Earth to make the calculation, because letting the Earth happen *was* the most efficient calculator. And letting the Universe happen is the most efficient calculation of everything, which of course is why the Universe has happened. The Universe is its own calculator.

So how is the scientist to make any worthwhile progress in our second category of many-particle problems? Progress, often very substantial progress, is made by two devices. One device is to degrade the basic physical laws into forms which permit calculations to be made much more simply and quickly than would otherwise be possible. Fortunately, the way things are, degradation often loses little and gains much. Most of atomic physics, of chemistry and biochemistry, can be calculated without bothering with the fine details of nuclear physics. And sometimes remarkable phenomena, like why a solid decides to melt at a precisely determined temperature, can be understood through extremely drastic degradations of the basic laws. In this last problem, it turns out that keeping a close eye on particle-to-particle relationships is more important than worrying about the individual particles themselves, which permits the properties of the individual particles to be vastly simplified, much more so than one could do in the study of chemistry or biochemistry.

The second sovereign method for achieving progress in complex many-particle problems is to use the results of experiments and observations, which is to say to use the outcome of calculations that the world performs on our behalf. The study of science is therefore a mosaic built from three components:

(1) Calculations from the basic physical laws, when such calculations are possible in simple cases.
(2) Calculations from degraded laws in more complex cases.
(3) Experiments and observations.

Of these, only (3) is consistently accurate. The basic physical laws have never been fully known, so (1) has never matched (3), although in the restricted range of what is known as 'quantum electrodynamics' (1) is believed with considerable justification to be nowadays as accurate as (3).

Laymen often find it difficult to understand why scientists are always in controversy with each other, and yet there never seems to be any room for the serious discussion of ideas which they themselves may put forward. The ideas which laymen put forward are almost always very simple, ideas like 'what happens if the speed of light varies', and they are readily dealt with by (1) alone. Nor do they involve issues where the basic laws are themselves in any real doubt. So the layman's challenge is usually over issues that can be calculated precisely and where the outcome of calculation is well-known. True scientific controversies, on the other hand, involve the details of the mosaic into which all of (1), (2), and (3) are to be fitted.

The causes of controversy lie mostly in the balance of (2) and (3), and scarcely ever in (1). A clash between (1) and (3) produces shock and bewilderment, not controversy. Shock and bewilderment, because such a clash implies that the elusive basic laws are still not basic enough. Once the possibility of experimental error has been eliminated, (3) is always preferred to (1), which is why science is the true determinant of the modern human environment. If scientists were to prefer (1) and (2) to (3), they would soon become as ineffective in their influence on the world as priests, economists, and politicians.

There is a distinction between a clash and a question of balance. A clash is a clear-cut contradiction between a calculation performed for a precisely defined set-up and an experiment or observation made for exactly the same set-up. How much common salt can be dissolved in a litre of water at a temperature of 25°C? The set-up here is defined and a calculation of the amount of dissolved salt could be compared with an experimental determination of the amount. Questions of balance, on the other hand, involve set-ups that are not precisely known. The inveterate practice of scientists working in biology, geophysics, geology, and astronomy, is to sharpen imprecise set-ups into precisely defined ones, by making assumptions which appear on subjective grounds to be 'reasonable'. One scientist may sharpen an imprecise set-up so that a contradiction emerges between (3) and (1) or between (3) and (2). Another scientist may sharpen the same imprecise set-up in a different way that avoids contradiction. So do controversies arise.

Let me give an example that will assume importance in a later chapter. Basalt is a rock with a characteristic chemical make-up (particularly with an unusually high calcium oxide content). Extensive sheets of molten basaltic lava have emerged out of rifts in the Earth's surface crust—on this there is no controversy. Detached basaltic rocks with weights up to a few pounds have been picked up from several sites on the surface of the Moon. These rocks too must once have been melted. But where? The set-up is too imprecise for this question to be answered without a further sharpening of the situation. Sharpening could come from additional facts, such as the observation of lunar volcanoes or by a visit to an actual lava bed. But no such facts are available, because no current volcanic activity has been observed on the Moon, nor have visits to several of the lunar maria (which were thought to be lava flows) revealed any actual lava beds. Even so, lunar geologists unanimously assert that there has been melting inside the Moon, and that the basalt has emerged there, much as it has done here on the Earth. Yet so far as the actual set-up is concerned, the basalt could have reached the Moon's surface from outside, with the melting which the basalt has clearly experienced having occurred in some other place. Lunar geologists reject this form of sharpening of the imprecise set-up, because to them it appears 'unreasonable'. Others, noting the crater-strewn lunar landscape, might think it certain that large quantities of material did indeed fall onto the Moon from outside.

Chapter 1

The Spin
of the
Sun

The problem of the origin of the solar system has been a matter of vigorous controversy for well over a century. Laplace proposed that the planets and satellites were formed contemporaneously with the Sun, but his particular model of how this might have happened fell into later disrepute, because the model turned out to be in conflict with the very principles of dynamics which Laplace had himself expounded so skilfully in his book *Mécanique Celeste*. But after flirting in the early decades of the present century with other ideas for the origin of the planets, astronomers are today more or less unanimous in thinking that Laplace was right in attempting to form both the Sun and planets in a single coherent process. The big issue of course is what process?

One school of thought, followed notably by A. G. W. Cameron and his colleagues, starts with an interstellar cloud of gas and with a condensation in the cloud. The procedure is then to follow the subsequent evolution of the condensation by calculation. Another school, to which I belong myself, is to start with the observed features of the solar system itself. This other approach, depending more heavily on observation than calculation, has an important advantage and an important disadvantage. The advantage is that the basis of the argument must be true. The disadvantage is that arguments going from effects to causes are not always unique. There may be more than one way of producing an observed situation, just as there is more than one way in which a piece of basaltic rock could conceivably arrive at the

Moon's surface. To defend against this difficulty one naturally seeks for the convergence of many facts toward a particular conclusion. Much of this book will be concerned with what many facts can be construed to tell us about the origin of the solar system.

I will begin with what I take to be the most remarkable fact of all, the slow spin of the Sun. Slow compared to what? To what we might expect for a blob of gas condensing from an interstellar cloud. Once a condensation becomes controlled by its own gravitation, the angular momentum stays constant, which means that the speed of rotation decreases inversely as the diameter, so that the period of the rotation decreases as the square of the diameter. Hence the expected final rotation period must satisfy the following order of magnitude condition,

$$\frac{\text{Initial period of rotation}}{\text{Final period of rotation}} \simeq \left(\frac{\text{Initial Diameter}}{\text{Final Diameter}}\right)^2 .$$

We know the present-day diameter of the Sun, 1.39×10^{11} cm, and for the initial diameter we can argue in this following way. Most stars are born in clusters, from which they escape eventually as the clusters gradually break up under the disruptive gravitational influence of our whole galaxy. Typically, a cluster has a diameter of about 15 light years, 1.42×10^{19} cm. The cluster in which the solar system was born may originally have contained 1,000 stars, with an average spacing from one star to another of about 10^{18} cm. The initial diameter to be inserted in the above formula cannot therefore be as large as 10^{18} cm, since otherwise adjacent stellar condensations would overlap each other. A reasonable estimate for the initial diameter would be an order of magnitude less, say 10^{17} cm, in which case the right-hand side of the above formula is 5.18×10^{11}. A rather more sophisticated way of arriving at this same estimate for the initial diameter goes as follows. There are several thousand clouds of gas in our galaxy in which stars are forming at the present time, with observed densities of hydrogen molecules up to 10^6 cm^{-3}. A sphere of such material with diameter 10^{17} cm would have about the mass of the Sun, 1.989×10^{33} g.

The initial period of rotation of a stellar condensation can scarcely be longer than the rotation period of the galaxy itself, about 2×10^8 years. Clouds of gas forming into stars rotate faster than this, and indeed observations of such clouds usually yield rotation periods of about 3×10^7 years, which I will use for the numerator on the left-hand

side of our formula. We then get

$$\text{Final period of rotation} = \frac{3 \times 10^7}{5.18 \times 10^{11}} \text{ years}$$

$$= 0.0212 \text{ days.}$$

The actual rotation period of the Sun is 25.3 days. The Sun spins very slowly compared to what we would expect.

The actual rotation speed of the Sun at its equator is close to 2 km s^{-1}. Our calculation would therefore indicate a rotation speed greater than this by the factor (25.3/0.0212), a rotation speed of $2.4 \times 10^3 \text{ km s}^{-1}$, and the Sun could not in fact spin as fast as that. If the actual Sun were made to spin faster and faster, it would become unstable at a rotational speed of about 400 km s^{-1}. Yet our calculation has indicated a speed about six times faster than this, which means that the solar condensation, or solar nebula as it is usually called, could not shrink to its present diameter—rotational forces would become dominant at a much larger diameter, in fact at 6^2 times the present diameter of the Sun (since the centrifugal force due to rotation is proportional to the *square* of the rotational speed). The solar nebula would then have a radius of about 2.5×10^{12} cm, about two-fifths of the radius of the orbit of the planet Mercury. As the solar nebula shrank towards this critical diameter it would become more and more flattened, with its equatorial diameter (the equatorial diameter is the one of relevance in the above calculation) about twice the polar diameter. At this stage the main body of the nebula would assume a lenticular shape and a disk of gas would emerge from the equatorial zone. Figure 1.1 is a schematic

Figure 1.1. During the condensation of the solar nebula a stage was reached when rotation had become fast enough for a disk of material to separate away at the equator.

representation of this situation, a situation that I will take no further in the present chapter, but which will be the starting point of Chapter 3.

To end this first chapter I would like to extend the above discussion to stars other than the Sun. The density of the gas in the interstellar clouds is by no means uniform, so that if we use the same initial diameter for the condensations that went to form other stars, the masses of the stars must inevitably vary from one to another. Since the final diameter of a condensed star varies with the mass M about as $M^{2/3}$, it follows that the right-hand side of the formula that we used above should be multiplied for other stars by $M^{-4/3}$ (when we measure M in terms of the Sun as the unit of mass). This means that the calculated final rotation period becomes longer as the mass M increases. Hence stars of large mass are better able than the Sun to withstand the forces generated by rotation. And not only are the rotational forces weaker because of the larger final diameters of more massive stars, but gravity is stronger for such stars, by the mass M itself. The combined effect of increased diameter and increased mass is therefore to give stars of large mass an advantage by $M^{5/3}$ in the containment of their rotational forces. This factor $M^{5/3}$ becomes greater than 6^2 for masses above about $10 \, M_\odot$ (where $M_\odot$ is the mass of the Sun). Such stars would be highly luminous, and are those which astronomers classify as of types B and O. Thus B and O stars may well be able to escape from the rotational crises that must overtake stars of lower mass like the Sun. They arrive at their larger final radii with rotational speeds that are less by $M^{-2/3}$ than the $2.4 \times 10^3 \, \mathrm{km \, s^{-1}}$ that was calculated above. For a star of mass $20 \, M_\odot$ the rotational speed would thus be about $325 \, \mathrm{km \, s^{-1}}$, a result that accords well with observed speeds for massive stars. This correspondence with observation may give us some confidence that our calculation for the solar nebula was correct, at any rate in its general features. In the next chapter we shall consider a quite different argument which points to just the same picture, namely that the solar nebula encountered the situation of Figure 1.1 when its equatorial diameter had decreased within the orbit of the planet Mercury.

Chapter 2

The Angular Momenta of the Planets

The planets lie far out from the present-day Sun. I do not recall ever seeing a diagram of the planetary orbits that did not suggest otherwise, since all such diagrams, merely from the thickness of the lines used to draw the orbits and from the dots used to denote the Sun and planets, give exactly the opposite impression, of the planets lying close by the Sun, in a kind of cosy relationship. I have never found a way to express the real situation better than an analogy I used many years ago:

'... let us represent the Sun as a ball 6 inches in diameter, the sort of thing you could easily hold in one hand .... Now how far away are the planets from our ball? Not a few feet or 1 or 2 yards as many people seem to imagine ... but very much more. Mercury is about 7 yards away, Venus about 13 yards away, the Earth 18 yards away, Mars 27 yards, Jupiter 90 yards, Saturn 170 yards, Uranus about 350 yards, Neptune 540 yards, and Pluto 710 yards. On this scale the Earth is represented by a speck of dust and the nearest stars are about 2,000 miles away'.

A more technical way of expressing the same thing is to say that the planets, in spite of their small masses compared to the stars, have remarkably large angular momenta. (The angular momentum of a particle of mass m, moving with speed v in a circle of radius r when taken about the centre of the circle is mvr. If the particle is moving in a non-circular orbit, then the speed v is replaced by the transverse

Table 2.1. *Planetary data*

Planet	Half of long axis of orbit (km)	Eccentricity	Average orbital speed (km s^{-1})	Time around orbit (years)	Mass (g)	Mass Earth masses	Radius (km)	Density (g cm^{-3})	Axial rotation period	Tilt of rotation axis to orbital plane
Mercury	5.791×10^7	0.206	47.90	0.2408	3.3×10^{26}	0.056	2439	5.4	58.7 days	7°
Venus	1.082×10^8	0.007	35.05	0.6152	4.9×10^{27}	0.81	6050	5.1	243 days	174°
Earth	1.496×10^8	0.017	29.80	1.0000	6.0×10^{27}	1.00	6378	5.52	$23^h 56^m$	23.5°
Mars	2.279×10^8	0.093	24.14	1.8809	6.4×10^{26}	0.11	3394	3.97	$24^h 37^m$	24°
Jupiter	7.783×10^8	0.048	13.06	11.8622	1.9×10^{30}	318	71,880	1.33	$9^h 55^{m\ a}$	3°
Saturn	1.427×10^9	0.056	9.65	29.4577	5.7×10^{29}	95	60,400	0.68	$10^h 38^{m\ a}$	27°
Uranus	2.869×10^9	0.047	6.80	84.013	8.8×10^{28}	15	23,540	1.60	$10^h 49^m$	98°
Neptune	4.498×10^9	0.008	5.43	164.79	1.0×10^{29}	17	24,600	1.6	15^h	29°
Pluto	5.990×10^9	0.249	4.74	248.4	—	—	—	—	—	—

[a] Temperate zones
[b] Data for Pluto for last six columns is uncertain.

component of the velocity.) More explicitly still, the angular momentum of each planet taken about the Sun can be calculated from the numerical data given in Table 2.1. The angular momenta of the larger outer planets, Jupiter, Saturn, Uranus, and Neptune, are indeed astonishingly large, much larger than the angular momentum possessed by the Sun itself. And this is in spite of the vastly greater mass of the Sun, 1.989×10^{33} g, more than a thousand times the mass of Jupiter, the largest planet. Thus the angular momentum of the Sun is about 1.7×10^{48} units (mass in grams, distance in centimetres, time in seconds), whereas the angular momentum of Jupiter is 1.9×10^{50} units. It follows therefore that the planets have a much greater share than the Sun of the angular momentum of the original solar nebula.

Suppose we argue that the material that went to form the planets acquired essentially the whole of the angular momentum of the solar nebula. Can we then, simply from the known planetary data (Table 2.1), work out the amount of angular momentum which the solar nebula must have had? The answer to this question is affirmative provided we take care over one very important point—there may have been more planetary material than we now see in the present-day planets. Indeed there must have been more, as we can decide with considerable assurance from the following argument.

Table 2.2 gives an extract of data from Table 2.1, together with a column headed 'Main constituents'. The latter column has been inferred, partly from the density values of the third column and partly from our knowledge of the main constituents of the Sun. The latter are given in Table 2.3, and the data of this last table is surely applicable also to the solar nebula, because the fractions given in Table 2.3 cannot have changed appreciably over the lifetime of the Sun.

Table 2.2. Density and main constituents of the major planets

Planet	Mass (Earth as unit)	Density $(g \, cm^{-3})$	Main constituents
Jupiter	318.00	1.33	Hydrogen and helium
Saturn	95.22	0.69	Hydrogen and helium
Uranus	14.55	1.6	Water, carbon dioxide, ammonia, methane
Neptune	17.23	1.6	Water, carbon dioxide, ammonia, methane

11

Table 2.3. Relative abundances of the commonest elements in the solar nebula

Element	Relative abundance (by number)	Fraction of total mass
H	3.18×10^{10} ⎫	0.9800
He	2.21×10^{9} ⎭	
C	1.18×10^{7} ⎫	
N	3.64×10^{6} ⎬	0.0133
O	2.21×10^{7} ⎭	
Ne	3.44×10^{6}	0.0017
Na	6×10^{4} ⎫	
Mg	1.06×10^{6} ⎪	
Al	8.5×10^{5} ⎪	
Si	10^{6} (standard) ⎬	0.00365
S	5×10^{5} ⎪	
Ca	7.2×10^{4} ⎪	
Fe	8.3×10^{5} ⎪	
Ni	4.8×10^{4} ⎭	

Of all the planets, Jupiter and Saturn are the only ones that possess hydrogen and helium in much the same high concentrations that the Sun does, the same high concentrations that existed in the solar nebula. Under terrestrial conditions the densities of hydrogen and helium are far lower than the densities of Jupiter and Saturn, but at the very high pressures within these planets the hydrogen and helium is much compressed. Indeed, calculations have shown that the expected densities of the high pressure states of these elements accord very well with a density of $1.33 \ \mathrm{g \ cm^{-3}}$ for Jupiter and $0.69 \ \mathrm{g \ cm^{-3}}$ for Saturn, the higher density of Jupiter being simply due to its larger mass, which compresses the hydrogen and helium more than the lower pressure within Saturn.

Pressures within Uranus and Neptune are significantly lower than within Jupiter and Saturn, too low to compress hydrogen and helium to densities anywhere near the observed values of about $1.6 \ \mathrm{g \ cm^{-3}}$. What then are the main constituents of Uranus and Neptune? In attempting to answer this question we naturally turn to the next most abundant elements, to the group CNONe of Table 2.3. In combination with hydrogen, and with each other, the CNO elements can form a whole range of molecules, H_2O, NH_3, CH_4, CN, CO, CO_2, and N_2. At the pressures within the solar nebula, H_2O, CO_2, and N_2 would be the most common of these, and at the higher pressures in Uranus and Neptune, CH_4 and NH_3 also form in appreciable quantities. Such materials within Uranus and Neptune would take on densities of about $1.6\,g\,cm^{-3}$. It is therefore reasonable to suppose that these two planets were constructed largely from the second most common group of elements in the solar nebula, namely from CNO, and Ne also perhaps, with the CNO in the forms of water, carbon dioxide, methane, and ammonia. But this second most common group of elements comprises only about 1.4 per cent of the solar nebula, so that the amount of material from which the water, carbon dioxide, methane, and ammonia must have been segregated had to have been some 70 times the sum of the masses of Uranus and Neptune, i.e. 70×31.78 Earth masses, or about 2,200 Earth masses.

To be a little more conservative, we should also take account of the third most abundant group of elements, Na, Mg, Al, Si, S, Ca, Fe, Ni, given in Table 2.3. Addition of these elements to Uranus and Neptune would reduce the needed amount of CNONe from the 31.78 Earth masses used in the preceding paragraph to a quantity somewhat less than this. It will be consistent with the more precise estimate given later in Chapter 5 if we take Uranus and Neptune to be made up from 25 Earth masses of CNONe with the remaining 6.78 Earth masses coming from Na, Mg, Al, Si, S, Ca, Fe, Ni, the third most abundant group of elements. Then we have $70 \times 25 = 1,750$ Earth masses as the amount of material whose segregation led to the formation of Uranus and Neptune.

We are therefore led to this remarkable conclusion: the original mass of the material from which all the planets in the solar system were formed must have been substantially greater than the present-day total mass of the planets, which is only about 450 Earth masses. Originally, the amount of planetary material must have been at least 1,750 Earth masses, the main fraction of which must have been lost in some way. The lost material consisted of some 1,300 Earth masses of hydrogen and helium. How did this loss occur?

The answer is rather obvious: through the evaporation of hydrogen and helium from the region of Uranus and Neptune. These regions are on the outskirts of the solar system, where the gravitational pull of the Sun is least able to restrain light gases like hydrogen and helium from streaming away into space. The gravitational pull of the Sun in the

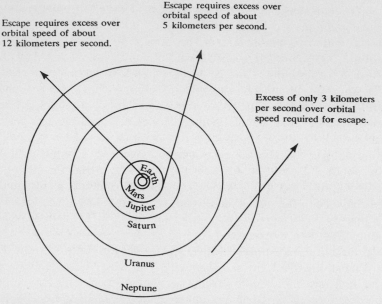

Escape requires excess over orbital speed of about 12 kilometers per second.

Escape requires excess over orbital speed of about 5 kilometers per second.

Excess of only 3 kilometers per second over orbital speed required for escape.

Figure 2.1. The gravitational pull of the Sun is stronger in the region of the terrestrial planets than it is in the region of Uranus and Neptune. The light gases, hydrogen, and helium, can escape most readily therefore from the outer part of the solar system.

region of Jupiter and Saturn is stronger, and is still stronger in the inner region of Mars, Earth, Venus, and Mercury. The situation is illustrated in Figure 2.1. In this figure, the planetary material in the outer regions is taken to be diffusely distributed, the condensation of H_2O, CO_2, N_2, CH_4, and NH_3 into Uranus and Neptune occurring at a later stage. This point is important, because the hydrogen and helium must have escaped *before* the formation of Uranus and Neptune, otherwise these planets would simply have picked up the hydrogen and helium through their own gravitational pull, and would then have become monster planets like Jupiter but of still larger mass, nearly 1,000 Earth masses each if the hydrogen and helium were shared equally. In Chapter 9 we shall find that Uranus and Neptune did indeed take a long time (of the order of 300 million years) to aggregate, which provided an ample duration for the hydrogen and helium to evaporate. Also in a later chapter (Chapter 8) we shall discuss the heat that was required to give atoms of hydrogen and helium fast enough speeds to escape in the manner of Figure 2.1.

A calculation of the amount of angular momentum possessed by the original planetary material is now easily made. Most of the angular momentum was possessed by the 1,750 Earth masses, much of which escaped from the solar system in the manner of Figure 2.1. Thus 1,750 Earth masses is 1.05×10^{31} g, the radius of Neptune's orbit is 4.498×10^{14} cm, and the orbital speed at this radius is 5.43×10^5 cm s^{-1}. Multiplying these numbers gives 2.55×10^{51} units of angular momentum.

Now if we write $V(R)$ for the equatorial rotational velocity of the solar nebula when its equatorial radius was R, the angular momentum was $k^2 M_\odot RV$, where k^2 is a number depending on the density distribution within the nebula. For a uniform density within a sphere the number k^2 is 0.4. The present-day Sun has quite a strong density concentration toward its centre, which reduces k^2 to a present-day value of about 0.05. An intermediate value between 0.05 and 0.4, 0.1 say, would be a reasonable choice.* Hence with the angular momentum equal to 2.55×10^{51} units we have the following equation

$$0.1\, M_\odot RV \simeq 2.55 \times 10^{51}.$$

This equation shows that as R decreased, V increased like $1/R$, as we took to be the case in the previous chapter. And as we saw in the previous chapter, V did not go on increasing throughout the whole condensation of the Sun, because at a certain stage V became so large that the instability of Figure 1.1 occurred. Let us see if we can now use our new equation to obtain an estimate (entirely independent of Chapter 1) for the stage at which the situation of Figure 1.1 must have taken place.

The physical condition for a disk to emerge from out of a rapidly rotating nebula is that the centrifugal force, given by V^2/R, should become comparable to gravity, $GM_\odot/R^2$, where G is the constant of gravitation. The problem of the exact relation between V^2/R and $GM_\odot/R^2$ (at which a disk would emerge) was studied many years ago by Jeans. Using Jeans' estimate for the so-called Roche model, given on page 225 of his book† *Astronomy and Cosmogony* one obtains the following equation,

$$\frac{V^2}{R} \simeq 1.5 \times 0.36075 \times \frac{GM_\odot}{R^2}$$

---

* This choice takes into account the flattened shape of the solar nebula. Without flattening, a choice of 0.2 would be better.

† Cambridge, 1928.

By eliminating V between our two equations we obtain

$$R = \frac{1}{GM_\odot} \left(\frac{2.55 \times 10^{52}}{M_\odot} \right)^2 \frac{1}{1.5 \times 0.36075}.$$

Inserting the known values of G and $M_\odot$, the result is $R = 2.3 \times 10^{12}$ cm, i.e. about two-fifths of the radius of the orbit of Mercury.

The correspondence of this calculation with the estimate obtained by the quite different argument in Chapter 1 is remarkable. It can give us confidence that the reasoning behind the estimates is substantially correct, and that the solar nebula encountered the rotational problem illustrated in Figure 1.1 when it had condensed inside the orbit of Mercury, to a radius $R = 2.3 \times 10^{12}$ cm according to our present discussion.

Chapter 3

Transfer of Angular Momentum

We return now to the stage reached at the end of Chapter 1, to the situation illustrated in Figure 1.1, with the rotation of the solar nebula so rapid that a disk of gas emerged from the equatorial zone (the stage at which $R = 2.3 \times 10^{12}$ cm, calculated in Chapter 2). Two arguments can be followed from this point onward. Although the arguments lead to very different conclusions, both seem to have correspondence with observed situations.

For the first argument we consider nothing but the properties of a gas, without the magnetic field that will appear in the second argument. Each unit mass of material in the disk of Figure 1.1 has more angular momentum than the average for the solar nebula as a whole, just because the disk has emerged from the equatorial zone where the angular momentum per unit mass is at its highest. Hence the effect of growing a disk is to reduce the average angular momentum per unit mass within the main body of the solar nebula. This has the effect in the main nebula of giving temporary relief from the centrifugal force caused by the rotation. But as shrinkage continues the main condensation spins up again, and the centrifugal force reasserts itself. Once again the main nebula is required to lose material into the surrounding disk, which therefore continues to grow as the inner body shrinks. So the disk becomes both massive and extensive, as shown in Figure 3.1. Calculation shows that about $\frac{1}{3}$ of the whole initial mass of the solar nebula would thus come to reside in the disk. Such a disk has no

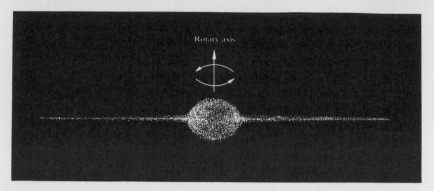

Figure 3.1. Without transfer of angular momentum from the central solar condensation to the surrounding disk the mass of material in the disk would become of stellar, not planetary, order.

correspondence with what we observe in the solar system. Even allowing generously for evaporation of hydrogen and helium the mass is much too large. Moreover, the disk would lie within the orbit of Mercury, instead of being outside it—as we observe the planets to be. Such a disk would eventually form itself into a second *star*, with a separation from the centre of mass of the system equal to about $\frac{1}{4}$ of the original radius of the disk, i.e. $\frac{1}{4}$ of 2.3×10^{12} cm $= 5.75 \times 10^{11}$ cm. The component stars in such a system would have a mass ratio of about 2 to 1, they would remain in rapid rotation, and their separation would be only a few times the sum of their diameters. Indeed they would be the kind of system known to astronomers as close binary stars.

To begin the second argument imagine the following experiment. A heavy uniform spheroid is made to spin freely about its polar axis. A lighter concentric circular ring in the equatorial plane is mounted so that it can turn freely about the same axis. Initially the ring has no rotation. Brushes are attached around the equatorial zone of the spheroid. They are made to extend outward so as to touch gently against the ring. What happens? Rather obviously, the brushes cause the ring to gain angular momentum at the expense of the spheroid. The latter slows its spin a little as the ring picks up speed, and this goes on until the spheroid and the ring go around together in the same period of rotation.

Now imagine the spheroid to be replaced by the main solar nebula and the lighter ring replaced by the disk which has just emerged (as in Figure 1.1) from the equatorial zone of the nebula. In this situation the disk and the main nebula have the same period of rotation—as the spheroid and the ring had at the end of our experiment. But as the solar nebula shrinks it spins up. If we consider 'brushes' of some kind to connect it with the disk, this spin-up will cause angular momentum to pass from the main nebula to the disk, as it did in the experiment. But whereas the transfer of angular momentum caused the ring to rotate faster until its period of rotation equalled that of the spheroid, something different happens for the solar nebula and disk.

To understand what this difference might be, consider the now familiar situation of a satellite moving in an orbit around the Earth. Suppose the satellite to be fitted with a rocket that is fired continuously with its jet always pointed opposite to the orbital motion. The effect would be to cause the satellite to spiral outward from the Earth, and as its distance from the Earth increased the satellite would actually *go more slowly*. As it gained angular momentum through the firing of the rocket, the period of the satellite around the Earth would lengthen, not shorten. So it would also be for the disk moving around the solar nebula. As it gained angular momentum due to the effect of the 'brushes', the disk would move outward, and its period of revolution around the main solar nebula would lengthen. Hence the main nebula and the disk experience opposite trends. Continued condensation causes the period of rotation of the main nebula to shorten, while angular momentum transfer causes the period of revolution of the disk to lengthen. Instead of coming to equality with each other, as in the mechanical experiment, the difference between the two periods tends to widen, and this keeps the process of angular momentum transfer in continued operation. Indeed the process can in principle take away essentially all of the angular momentum of the main nebula, passing it out to the disk; and this is just what must have taken place during the formation of the solar system. A process of this kind is essential if we are to understand how the Sun came to spin so slowly, and why the angular momenta of the planets are so large.

Unlike the first argument given above, this second argument does not require material to be added continuously to the disk. Provided the process of angular momentum transfer is strong enough, i.e. provided the 'brushes' are firm enough, there need be no further additions of material to the disk after its original emergence in the situation of Figure 1.1. The necessary condition is that transfer be more rapid than the shrinkage of the main nebula. At a radius of 2.3×10^{12} cm the time-scale for shrinkage can be calculated to be about 4,000 years (as we shall see in Chapter 4). The 'brushes' must be firm enough to operate in a time-scale less than this, if the mass in the disk is to be kept down to planetary order—i.e. to about 1 per cent of the mass of the main nebula—instead of becoming of stellar order as it did inevitably in the first argument.

The essential property of our 'brushes' is that they serve as a torque transmitter, reacting back on the solar nebula to take away angular momentum, as well as acting on the material of the disk. The torque transmitter for the solar nebula was probably magnetic in its nature. The solar nebula must have possessed a dipolar magnetic field. A pulling-out of such a field, by material passing from the main nebula to the disk, produces a field whose intensity falls off inversely as the square of the distance from the nebula—not inversely as the cube of the distance, as with a purely dipole field.

Any form of pulled-out field has this inverse square property, and would serve as a 'brush' in the sense of the above argument. In order for the field to be firm enough to transfer most of the angular momentum from the solar nebula to the disk within a time-scale of 4,000 years, the magnetic intensity at the nebula must exceed about 100 G. Since magnetic intensities up to about 2,000 G are observed in present-day sunspots, this requirement does not seem at all unreasonable.

There is nothing in principle to stop a torque transmitter from acting across a complete vacuum. The tides which the Moon raises in the waters of the terrestrial oceans causes friction along the land margins, and this friction creates a torque transmitter between the Earth and the Moon that would operate just as effectively even if there were a complete vacuum between the Earth and the Moon. A magnetic torque transmitter based on a purely dipole field could also act across a vacuum, but a magnetic torque transmitter based on a pulled-out field demands at least one sheet of conducting ionized gas to exist within the interspace.

This is because a pulled-out field has at least one surface at which the magnetic component parallel to the surface undergoes a sudden reversal, and an electric current must flow in a surface sheet of conducting material to maintain this reversal. For a pulled-out dipole field there is one such surface, and it is the equatorial plane. Not a great deal of gaseous material is required because the sheet can be quite thin. The conduction of the electric current by free electrons is self-maintaining, since so-called eddy currents would immediately be set up should too many electrons recombine with positive ions (forming neutral atoms); and the eddy currents create voltages which ionize more of the atoms. Moreover, the thickness of the sheet adjusts itself so that the pressure within the gas corresponds appropriately to the pressure which the magnetic field exerts upon the sheet.

Finally in this chapter we note that the establishment of magnetic 'brushes' between the solar nebula and the outer disk of planetary material need not be a one-shot affair. If some instability in the conducting sheet were to drain away the ionized gas, the pulled-out property of the field would disappear, but this could be re-established through particles being ejected by the solar nebula in the manner of the present-day solar wind. Such particles would carry a pulled-out magnetic field back again to the planetary disk. In this connection it is satisfactory that observations of newly-formed stars show them to be highly disturbed at their surfaces, and it is thought that magnetic fields play an essential role in maintaining their intense activity.

Chapter 4

The Luminosity of the Solar Nebula

The surface temperature which the solar nebula would have had at the stage of Figure 1.1, the equatorial radius R of the nebula being then about 2.3×10^{12} cm, is rather well-known. It would be about 3,500 K. The reason for this temperature is quite strong, and it lies in an effect which Martin Schwarzschild and I discovered many years ago in the study of a different problem, the structure of giant stars.

The definition of the 'surface' of a star, or of a protostar like the solar nebula, requires the opacity of the surface material to be properly adjusted, so as to be just able to allow radiation to escape outward into space. If the opacity were too little, radiation would escape from lower levels within the star, and those lower levels would soon become the 'surface' themselves. If the opacity were too much, there could be no uniform outflow of energy through the star's interior, and there would be a general readjustment of either the subsurface layers or of the deeper structure of the star until the surface opacity was appropriately reduced.

What Schwarzschild and I found was that the surface material of a star cannot have sufficient opacity to meet this condition if its surface temperature drops much below 3,500 K. Stars with very large radii, giant stars, can have somewhat lower temperatures, into the range 2,500–3,000 K, but this is about the least boundary temperature that a true star (not simply a diffuse cloud of gas) can have.

The Japanese astronomer, C. Hayashi, applied this opacity condition to newly-forming stars like the solar nebula. What Hayashi found was that for newly-forming stars the ratio of the density of material at the centre to that at the surface was unusually low, so low indeed that newly-forming stars must be convective throughout their interiors. Indeed, the activity of the internal convection adjusts itself so as to get the opacity condition of the surface exactly right. When it does so, the surface temperature is about 3,500 K.

This result permits the luminosity of the solar nebula to be easily estimated from a simple comparison with the present-day Sun which radiates at a power output of 3.8×10^{23} kW (1 kilowatt $= 10^{10}$ ergs per second). The emission from each unit area of a hot surface is proportional to the fourth power of the temperature, and this requires the emission from unit area of the surface of the solar nebula to be less than that from the present-day Sun by the factor $(3,500/5,700)^4$, because the surface temperature of the present-day Sun is about 5,700 K. But the surface of the solar nebula had a larger number of unit areas than the present-day Sun has, by approximately the factor $(R/R_\odot)^2$, where R was the equatorial radius of the solar nebula and $R_\odot$ is the radius of the present-day Sun. (This factor is approximate since it assumes the surface of the solar nebula to have been a sphere.) To sufficient accuracy the luminosity of the solar nebula was therefore

$$\left(\frac{3,500}{5,700}\right)^4 \times \left(\frac{R}{R_\odot}\right)^2 L_\odot,$$

where $L_\odot$ denotes the luminosity of the present-day Sun, 3.8×10^{23} kW. Putting $R_\odot = 6.96 \times 10^{10}$ cm, and using the magic number $R = 2.3 \times 10^{12}$ cm that we calculated in Chapter 2, the luminosity of the solar nebula at the emergence of the disk of planetary material (Figure 1.1) comes out to be about 150 $L_\odot$. This very high luminosity would be maintained through the main phase of the torque transmission that was discussed in Chapter 3. *Thus the first stages in the history of the formation of the planets was enacted against the background of a very high luminosity from the main central nebula.*

The interior of the solar nebula was not deriving any significant quantity of energy from nuclear processes at the stage of Figure 1.1. Thus the time required for shrinkage to occur from this stage was determined by the ratio of the gravitational energy to the luminosity. The gravitational energy was less than that of the present-day Sun by approximately the factor $R_\odot/R$, so that the shrinkage time was approximately

$$\frac{\dfrac{R_\odot}{R} \times (\text{Gravitational Energy})_\odot}{\left(\dfrac{3,500}{5,700}\right)^4 \left(\dfrac{R}{R_\odot}\right)^2 L_\odot},$$

where (Gravitational Energy)$_\odot$ is that of the present-day Sun. Now (Gravitational Energy)$_\odot/L_\odot$ was calculated almost a hundred years ago by Lord Kelvin, who found it to be about 2×10^7 years. Inserting Lord Kelvin's result gives $(R_\odot/R)^3(5,700/3,500)^4(2 \times 10^7)$ years for the shrinkage time from the stage of Figure 1.1. Again putting $R_\odot = 6.96 \times 10^{10}$ cm, $R = 2.3 \times 10^{12}$ cm, we arrive at an estimate of about 4,000 years, a result mentioned already in Chapter 3.

As the solar nebula condensed from $R = 2.3 \times 10^{12}$ cm to smaller radii, the luminosity declined proportionately to R^2 and the shrinkage time increased as $1/R^3$. Thus at $R = 2.3 \times 10^{11}$ cm the luminosity had declined to a value that was comparable to the present-day Sun, and the condensation time had lengthened to several million years (not exactly the 4 million years that one would calculate from the above formula, because the surface temperature of the nebula would have increased moderately from 3,500 K to about 4,000 K). With R decreasing still further towards $R_\odot = 6.96 \times 10^{10}$ cm, the luminosity of the solar nebula actually dipped below $L_\odot$, so that the early very high luminosity phase was followed several million years later by a low luminosity phase.

The final stages of condensation, with R decreasing below 2.3×10^{11} cm, cannot be followed with sufficient accuracy from the above argument, however, because a crucially new physical situation then arose in the deep interior of the primordial Sun. Nuclear reactions began to generate significant quantities of energy, through the transformation of hydrogen into helium. Much more ambitious calculations are needed to follow these final stages in detail. From such calculations it turns out that the luminosity of the primordial Sun rose from its low dip, but *not* to its present-day value, $L_\odot$. The luminosity rose to about seven-tenths of the present-day value, $0.7 L_\odot$.

The generation of energy by nuclear processes stopped the shrinkage of the primordial Sun. The solar nebula had become a true star at last, with a lifetime ahead of it to be measured, not in thousands or even in millions of years, but in thousands of millions of years. By the time the

Sun had become a true star the stage for the formation of the planets had already been set. It had been set when the radius of the solar nebula was about 2.3×10^{12} cm, when the luminosity was more than a hundred times greater than the present-day luminosity. The remarkable consequences of this very high luminosity of the solar nebula will emerge in the next two chapters.

Chapter 5

Segregation by Condensation

As the ring of planet-forming gas moved outward from the solar nebula in the manner discussed in Chapter 3, the temperature of the gas must have declined approximately as the inverse square root of the radius r of the ring. Indeed, the temperature would be given by the simple formula

$$\sim 3{,}500 \left(\frac{R}{r}\right)^{1/2} K,$$

R being the radius of the solar nebula, $R \simeq 2.3 \times 10^{12}$ cm according to Chapter 2. This leads to the temperature values set out in the third column of Table 5.1. These values have been calculated at values of r equal to the radii of the planetary orbits (the unit of r in Table 5.1 is the 'astronomical unit', which is the radius of the Earth's orbit, 1.496×10^{13} cm).

As the temperature of the planetary material declined solid and liquid particles would condense, thereby segregating refractory substances, like Fe, MgO, SiO_2, Al_2O_3, from the more volatile materials, which still remained in gaseous form. We shall take a detailed look in the next chapter at the temperatures at which various substances condense. Here we simply note that 1,500 K is a typical temperature for the refractory group just mentioned. According to Table 5.1 this temperature fell about mid-way between the orbits of Venus and the Earth. More volatile elements like sodium and potassium condensed

Table 5.1. Temperature values in the ring of planetary gas

Planet	Radius r (AU)	Temperature (K)
Mercury	0.387	2,206
Venus	0.723	1,613
Earth	1.000	1,372
Mars	1.524	1,111
Jupiter	5.20	602
Saturn	9.54	444
Uranus	19.18	313
Neptune	30.07	250

out of the gas at about 1,100 K, which temperature fell in the region of the planet Mars, while H_2O, CO_2, N_2, require temperatures less than 200 K, and so the most volatile materials continued to remain gaseous out to the regions of Uranus and Neptune. Water, carbon dioxide and nitrogen would of course eventually condense when, with continuing shrinkage, the luminosity of the solar nebula declined from its very high value at $R = 2.3 \times 10^{12}$ cm.

The magnetic 'brushes' which transmitted the torque that drove out the ring of planetary gas did not act appreciably on solid or liquid particles, which would therefore remain at the values of r at which they condensed, unless the particles were carried along by viscous drag from the gas. Whether or not the particles were thus carried along by the gas depended on their sizes. Rather obviously, sufficiently fine particles would go with the gas, while big enough chunks of the material would not. Big enough chunks became segregated according to their condensation temperatures.

So far in this book we have been arguing in an inferential style, working from features of the solar system that we know to be true, features like the slow spin of the Sun, and the large angular momenta of the planets. But now we have arrived at a deductive result, a result that I have not seen come out in a plausible way in any other theory of the origin of the planets. Namely that the inner planets must be composed of refractory materials, and that the masses of the inner planets must be comparatively small, simply because the refractory materials were present in the planetary gases only in the low abundances of Table 2.3. And as a matter of detail we have even calculated the place of greatest congestion of these materials, mid-way between Earth and Venus, as they must surely have been.

But we have not yet escaped from this first confrontation of theory and observation with an entirely whole skin. According to Chapter 2 there were about 1,750 Earth masses of planetary gas, and according to Table 2.3 the main group of refractory elements, Na, Mg, Al, Si, Ca, Fe, and Ni, comprise a fraction 0.00365 of these 1,750 Earth masses. And if we allow for oxidation, with sodium, magnesium, aluminium, silicon, and calcium, in the forms Na_2O, MgO, Al_2O_3, SiO_2, and CaO, the fraction is increased to about 0.0050. There would thus be rather more than 8 Earth masses of refractory materials in the whole of the planetary gas, considerably more than the 2 Earth masses or so contained in the terrestrial planets.

The escape route from this situation is of course quite clear. We have to argue that only 2 Earth masses of the refractories condensed as particles that were big enough to withstand the viscous drag of the gas. The remaining 6 Earth masses went with the gas as it moved outward to the regions of Uranus and Neptune, making an eventual contribution to the masses of these outer planets (of which we took account in Chapter 2). The desirable position would be to verify this contention by actual observation, and certainly refractory materials are observed in the outer part of the solar system. The satellites Io and Europa of Jupiter have densities like the Moon and are therefore almost certainly composed of rock. And cometary material is thought to contain small refractory particles. These examples would require refractory materials to have been carried outward by the gas—but there is no observational requirement for the amount of the material so carried to have been nearly as large as we have estimated it to be.

To obtain further insight into the problem some fairly straightforward calculations can be made however. At the stage of importance for the inner planets, consider the gas to occupy a penny-shaped volume of radius 1 AU (1.496×10^{13} cm) and thickness 10^{12} cm. There being 1,750 Earth masses, the average density of the gas comes out to be about 1.5×10^{-8} g cm^{-3}. (At a temperature of 1,500 K, with molecular hydrogen the dominant component, the pressure would be close to 10^{-3} atm, a result that will be of interest in the next chapter.)

There is a device whereby a swarm of particles can co-operate together to resist the outward drag of the gas, and this is by falling under gravity toward the equatorial plane and forming a sheet there with a high density of refractory material. A particle condensing at a height z above the equatorial plane experiences an acceleration $GM_\odot z/r^3$ toward this plane, arising from the gravitational field of the main solar nebula, and the particle begins to fall toward the plane because of this acceleration. But as it picks up speed it encounters an increasing resistance from the gas. Eventually the speed rises to the point at which the resistance equals the gravitational force—the particle is then said to have reached its terminal velocity. The question now is whether the terminal speed is large enough for the particle to cover most of the distance z within the time-scale available, i.e. the time-scale for which the gas resides in the vicinity of the orbits of the inner planets, a time-scale of perhaps 1,000 years. The answer to this question turns on the size of the particle, since the larger and more massive the particle the higher is its terminal speed. At the gas density in question, $1.5 \times 10^{-8}\,\mathrm{g\,cm^{-3}}$, I find that a size of order 1 cm is necessary if a particle is to come close to the equatorial plane within a time of 1,000 years.

There is no contradiction in supposing that the main refractory materials could form particles as large as this. For iron, as an example, the density would be about $1.5 \times 10^{-11}\,\mathrm{g\,cm^{-3}}$. *If this density were maintained in the gas phase*, iron particles could grow (in 1,000 years) to sizes considerably larger than 1 cm, to sizes of some tens of metres. But there is a possible snag in the question: would there not be so many condensation nuclei that the available amount of iron vapour proved insufficient to enable more than a fraction of them to grow to metric sizes, perhaps indeed only a small fraction? This is a harder question than any we have yet encountered but the following consideration goes some way toward answering it.

If one looks at the temperature values in the third column of Table 5.1 it seems as though it must have been too hot in the region of the orbit of Mercury for any condensates to form there, and this would be true if these temperature values could be interpreted in the usual way. But the radiation from the solar nebula is directed outward, it does not come at the particles isotropically as it would in a strictly thermal situation. This leads to a situation in which a particle alone can be cooler than it would be if it were embedded in a swarm of other particles; because a particle alone receives radiation only from the main solar nebula, whereas in a multi-particle situation each particle also receives radiation from the other members of the swarm. The temperatures of Table 5.1 apply to the multi-particle case. A particle alone can be cooler than these values by the factor $1/\sqrt{2}$ (provided the material of the particle is a good enough conductor of heat for a uniform temperature to be established within it). For Mercury, this reduces the temperature from the value 2,206 K in Table 5.1 to about 1,500 K, which is certainly cool enough for Al_2O_3 to condense and is indeed not far from the condensation temperature of iron itself.

Only a small fraction of the iron vapour (for example) can condense at these lowered temperatures, because if too many particles were to form each would then receive radiation from the others, and temperatures would rise toward the values of Table 5.1, with the consequence that iron particles interior to the orbit of Venus would evaporate. So with an ample supply of vapour available the favoured few particles can grow to large sizes, and they can then resist the outward flow of the gas.

As the gas reached the region between the orbits of Venus and the Earth, condensates could form without specially lowered temperatures being necessary. Precipitation would now be very much heavier, with many particles competing for a limited supply of vapour. The particles would therefore be smaller, probably with most of them too small to fall to the equatorial plane, and so to resist the drag of the gas. The bulk of the mass of the condensates continued with the gas—we require about 1 part in 4 of it to be in the form of particles with sizes above a centimetre (that can approach the equatorial plane) and this does not seem an unreasonable proportion.

As a final remark in this chapter, it was noted above that in order for the $1/\sqrt{2}$ factor to be applied to the temperature values of Table 5.1 there must be a good heat conduction through a lone particle. For lone particles growing to large sizes, tens or hundreds of metres in diameter, metallic material would best meet this condition. Perhaps it is for this reason that the planet Mercury contains so much iron.*

---

* The average density of Mercury, 5.4 g cm^{-3}, demands a high proportion of a material with density above 5.4 g cm^{-3}, in order to compensate for the lower density of rock, which is about 3.2 g cm^{-3}. Iron is the commonest material with such a required density. The mass of Mercury is so low that compression of materials is not a major factor.

Chapter 6

Mineral Condensations in the Planetary Material

A few years ago Professor L. Grossman made calculations of mineral condensations in a cooling solar nebula.* His calculations are remarkably adaptable to the outward-moving planetary material described in previous chapters. In particular, the pressures of about 10^{-3} atm which the outward-moving gases had when they were in the region of the orbits of the inner planets was just the pressure at which Grossman computed most of his results. And whereas many of the mineral condensates at the higher temperatures became lost as the solar nebula cooled in Grossman's model, all of the condensates are preserved in some degree in the present model. Preservation comes about because some of the condensates (those falling to the equatorial plane) are left behind as the planetary gases move outward and their composition remains frozen (but see comment at the end of this chapter). Materials continuing with the gas are not frozen in composition, however. As the temperature within the gas declines they undergo transformations of mineral composition in just the way discussed by Grossman. These remarks will become clearer in meaning after the calculations themselves have been presented in graphical form.

Table 6.1 gives a list of the minerals that were considered in the calculations, while Figures 6.1 to 6.4 trace the thermal histories of the most abundant refractory materials, Al, Mg, Si, Ca, and Fe. The iron,

* L. Grossman, 'Condensation in the Primitive Solar Nebula', *Geochimica et Cosmochimica Acta*, **36** (1972), 597.

Table 6.1. Chemical formulae of minerals referred to in Grossman's paper

Solid solution series	Mineral	Formula
Olivine	Forsterite	Mg_2SiO_4
	Fayalite	Fe_2SiO_4
Pyroxene		
Orthorhombic	Enstatite	$MgSiO_3$
	Ferrosilite	$FeSiO_3$
Monoclinic	Diopside	$CaMgSi_2O_6$
	Tschermak's molecule	$CaAl_2SiO_6$
	Ti-Al-pyroxene	$CaTiAl_2O_6$
Plagioclase	Anorthite	$CaAl_2Si_2O_8$
feldspar	Albite	$NaAlSi_3O_8$
Melilite	Gehlenite	$Ca_2Al_2SiO_7$
	Akermanite	$Ca_2MgSi_2O_7$
	Soda-melilite	$CaNaAlSi_2O_7$
Nickel-iron		(Fe, Ni, Co, Cr)
	Corundum	Al_2O_3
	Perovskite	$CaTiO_3$
	Hibonite	$CaO \cdot 6Al_2O_3 \pm MgO, TiO_2$
	Spinel	$MgAl_2O_4$
	Magnetite	Fe_3O_4
	Schreibersite	$(Fe, Ni)_3P$
	Troilite	FeS
	Feldspathoids	
	Nepheline	$NaAlSiO_4$
	Sodalite	$3NaAlSiO_4 \cdot NaCl$
	Wollastonite	$CaSiO_3$
	Grossularite	$Ca_3Al_2Si_3O_{12}$
	Cordierite	$(Mg, Fe)_2Al_4Si_5O_{18}$

which condenses simply as metal, appears in the same figure as the magnesium. A line drawn parallel to the ordinate at any temperature, on any one of the figures, determines the fractions in various mineral forms of the element in question. These are the fractions which fall out of the gas at that particular temperature. For example, aluminium falls out of the gas as corundum at 1,700 K and as spinel at 1,400 K. Above 1,300 K magnesium falls out mostly as forsterite, but at 1,200 K mostly as enstatite. Here already we have a distinction between a magnesium-rich, ultrabasic rock (forsterite) and a rock with a 60 per cent silica content (enstatite). When we consider that the temperature

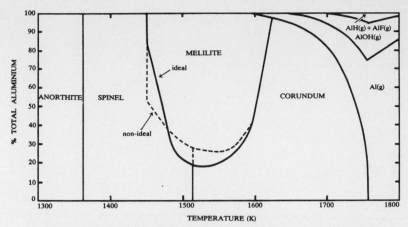

Figure 6.1. The distribution of Al between crystalline phases and vapour in a cooling gas of solar composition at 10^{-3} atm total pressure. Corundum, appearing at 1,758 K, is the first Al-bearing condensate. Melilite is seen to form by the reaction of corundum with the vapour, producing pure gehlenite at first. Below 1,550 K, the reaction of gehlenite with the vapour begins to increase the akermanite content of the melilite, reaching 81 mole % or 43 mole %, according to ideal and non-ideal solid solution models, respectively. The displaced Al forms corundum above 1,513 K and spinel below this temperature, at which corundum reacts with the vapour to form spinel. The reaction of melilite to form diopside and spinel at 1,450 K and the reaction of diopside and spinel to form anorthite at 1,362 K are also shown [from L. Grossman, 'Condensation in the Primitive Solar Nebula', *Geochimica et Cosmochimica Acta*, **36** (1972), 597].

difference between 1,300 K and 1,200 K implies a difference, in the manner of Table 5.1, of the positions in the solar system at which these two rock types fall out, it already becomes clear that by no means all of the differentiation of rock types took place after the Earth was formed. There was differentiation already in the beginning, and perhaps indeed the beginning was the most important of all differentiations.

It was explained already in Chapter 5 that the temperature values of Table 5.1 are not strictly thermodynamic. The temperature that a particular particle or a particular rock sample actually attains depends not only on its distance from the main solar nebula (which is the only variable in Table 5.1) but also on the relation of the sample to surrounding material. The temperature can be increased by radiation received from the surrounding material, in the way that radiation from buildings increases the temperature in a city on a warm summer day.

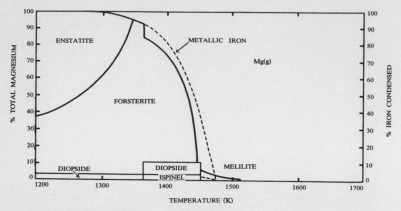

Figure 6.2. The distribution of Mg between crystalline phases and vapour in a cooling gas of solar composition at 10^{-3} atm total pressure. The condensation curve of metallic iron is also shown. Note that neither iron nor a significant amount of Mg condenses until well below the temperatures where Ca, Al, and Ti are 100% condensed. Before forsterite appears at 10^{-3} atm, 46% of the iron is condensed [from L. Grossman, 'Condensation in the Primitive Solar Nebula', *Geochimica et Cosmochimica Acta*, **36** (1972), 597].

Or one aggregation of material can shield another aggregation from the direct radiation of the solar nebula, giving the opposite effect of a lowering of temperature for the shielded aggregate. Indeed, in the complex situation near the equatorial plane there must certainly have been temperature fluctuations of several hundred degrees kelvin between material samples located at essentially the same distance from the solar nebula. And since the calculated condensation temperatures of quite a number of minerals are only a few hundred degrees kelvin below their melting temperatures it seems clear that rock samples would be melted from time to time. Changing configurations can cause alternating phases of melting and solidification, permitting particles to coagulate into rocks, and permitting mineral crystallization within the rocks. The high luminosity of the solar nebula must have been maintained for about 4,000 years, and in such a span of time many transformations which we normally ascribe to geochemistry could have taken place. The environment is very rich in its variables, particularly with respect to chemical composition and to fluctuations in temperature of several hundred degrees kelvin. At the outset we remarked on the possibility that pieces of basalt might have arrived at the Moon's

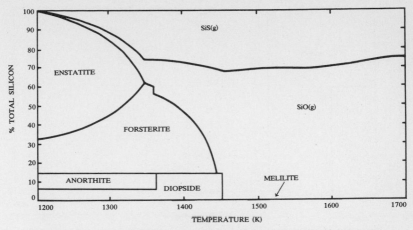

Figure 6.3. The distribution of silicon between crystalline phases and vapour. Melilite is the first silicon bearing phase to condense. The curve shown is for the case of ideal solid solution but the non-ideal curve is very close to this Less than 15 per cent of the total Si is condensed until forsterite appears at 1,444 K. The rate of condensation of Si with falling temperature is further increased below 1,349 K where forsterite begins to react with the gas to form enstatite [from L. Grossman, 'Condensation in the Primitive Solar Nebula', *Geochimica et Cosmochimica Acta*, **36** (1972), 597].

surface from outside. This possibility does not seem as contrived now as it may perhaps have done before.

There is an aspect to Figures 6.1 to 6.4 that requires some comment. I have referred to materials falling out of the gas as becoming frozen with the composition appropriate to the temperature of fall-out. There will, however, be a tendency for such materials to evaporate once the gases have gone. The calculations represented in Figure 6.2, say for a temperature of 1,400 K, require Mg to be present as a gas (and also Si, SiO). With the gases swept away, the solid materials evaporate toward a situation that renews the Mg gas. And if the Mg gas (and other gaseous components) are swept continuously away, evaporation continues, and we might wonder if the eventual effect might not be to dissipate all the previously condensed solid materials.

Evaporation would become complete if the temperature was maintained for long enough. But with a time scale of no more than 4,000 years for the high temperature phase, the larger bodies that aggregate together near to the equatorial plane will not undergo serious evaporation. It is possible that effective condensation temperatures may be

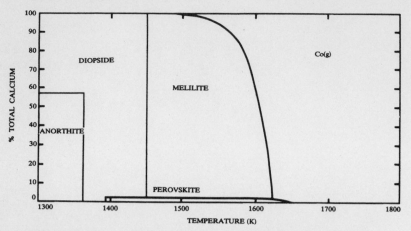

Figure 6.4. The distribution of Ca between crystalline phases and vapour in a cooling gas of solar composition at 10^{-3} atm total pressure. Ca first condenses as perovskite at 1,647 K, but the major sink for Ca at high temperature is melilite, which condenses at 1,625 K. The reactions of melilite to form diopside and spinel at 1,450 K and diopside plus spinel to form anorthite at 1,362 K are also shown [from L. Grossman, 'Condensation in the Primitive Solar Nebula', *Geochimica et Cosmochimica Acta*, **36** (1972), 597].

lowered somewhat due to this effect, possibly by as much as 100 K— lowering the temperature only slightly reduces evaporation rates very appreciably. There are, moreover, the effects discussed above that work in the opposite direction, the $1/\sqrt{2}$ effect of Chapter 5 and the effect of the shielding of one body by another. The determination of precise temperature values is evidently a highly complex problem, but so long as we are content with estimates that are good to within one or two hundred degrees, then the values given in Table 5.1 can be used, together with calculated fractionations shown in Figures 6.1 to 6.4.

Chapter 7

The Terrestrial Abundances of the Rare Gases

By now we have accumulated some problems. In particular, we have lost elements of low abundance from the region of the inner planets. Table 7.1 gives estimates of the condensation temperatures of the rarer refractory materials in the planetary gases. (For a table of the relative abundances of all the elements, as they are believed to have been in the solar nebula, see the Appendix.) The crystalline phases given in the second column of Table 7.1 could not, because of the low abundances of these elements, form themselves into particles that were large enough to fall out of the gas. Unless these rarer elements could substitute for a common element in a mineral which formed particles of large size, there is no question but that all these elements would be contained only in fine smoke particles, which would easily be carried along by the outward-moving gases. Uranium and thorium would be among the very first condensates, and being elements of very low abundance would be among the elements carried as fine particles along with the gases as they moved outward.

And of course we have lost from the region of the inner planets all the important volatiles, particularly all those molecules which involve H, C, N, and O, like H_2O, N_2, CO_2. The differentiation which has occurred appears at first sight to be too drastic. The position would really be disturbing if there were not a crucial observation to show that the rock and iron which forms the main body of the Earth were indeed once stripped almost entirely clean of all volatiles. The

Table 7.1. Condensation temperatures of re-fractory trace elements and those of the major high-temperature minerals[a]

Gaseous species	Crystalline phases	Condensation temp. $(10^{-3}$ atom)
ThO_2[b]	ThO_2	Above 2,000 K
UO_2[b]	UO_2	Above 2,000 K
Os	Os	1,925
W, Wo	W	1,885
ZrO_2, Zr, ZrO	ZrO_2	1,840
Re	Re	1,839
	Corundum (Al_2O_3)	1,758
HfO_2, HfO, Hf	HfO_2	1,719
ScO, Sc	Sc_2O_3	1,715
Mo, MoO	Mo	1,684
	Perovskite $(CaTiO_3)$	1,647
ReO, Re	Re_2O_3	1,647
Ir	Ir	1,629
	Gehlenite $(Ca_2Al_2SiO_7)$	1,625
Ru	Ru	1,614
VO, V	V_2O_3	1,534
TaO, Ta	Ta_2O_3	1,499

[a] Table adapted from L. Grossman and J. W. Larimer, 'Early Chemical History of the Solar System', *Reviews of Geophysics and Space Physics*, **12**, (1974) 71. The first column shows the gaseous molecules for each element that were included in the calculations, while the second column gives the highest-condensing phase.
[b] My estimates.

observation has three bits to it:

(1) The ratios of the masses of neon in the atmosphere to nitrogen in the atmosphere to oxygen in the ocean waters is $1:60,000:1.5\times10^7$, whereas the ratio in the solar nebula was (Table 2.3) $1:0.75:5$.

This enormous depletion of neon cannot be explained by thermal evaporation because.

(2) The ratios of the masses of neon, krypton, and xenon for the atmosphere are much the same as they are for the neon, krypton, and xenon occluded within the surface rocks of the

Earth. This is a strong indication that atmospheric Ne, Kr, Xe have been outgassed from the rocks. Then, since Xe has surely not experienced thermal evaporation because of its large atomic weight, Ne cannot have done so (otherwise Ne would be much depleted in the atmosphere relative to Xe, which it is not).

(3) Unless the rocks of the Earth's interior contain enormously more occluded Ne, Kr, Xe than the surface rocks, a total outgassing of the Earth would not increase the atmospheric concentrations of these gases by more than a moderate factor, a factor perhaps of about 7. The concentrations relative to nitrogen and oxygen would still be exceedingly low compared to the concentrations in the solar nebula.

The argument can be restated in a somewhat different way. Because the abundance of Xe is very low and because Xe cannot have been thermally evaporated there must have been a time on the Earth when all the volatiles were also in low abundance, so low indeed that most of the present-day volatiles must have been added subsequently. But why was water, carbon dioxide, and nitrogen added later and not Ne, Kr, Xe? To anticipate the discussion of a later chapter, because materials added later came as solid ices, which H_2O, CO_2, NH_3 could be, whereas Ne, Kr, Xe were gases and could not be added—the difference was one of phase.

Just as I regard the slow spin of the Sun as the crucial fact to be explained by a theory of the origin of the planets, so I regard the low Ne, Kr, Xe concentrations in the terrestrial atmosphere as the lynch pin of a theory of the detailed history of the Earth. To me, theories which offer contrived explanations of these facts, or which sweep them entirely under the rug, are not theories to be considered seriously.

The rare gases become occluded in rocks more than other volatiles, especially at high temperatures. The very low concentrations on the Earth represent in my view the maximum retention of volatiles by the terrestrial rocks. The rocks were not hydrated. Nor was there any retention of CO_2. Consequently the oceans and the CO_2 present nowadays in the limestone rocks have not come from outgassing of the Earth. In short, there was an era when the main constituents of the Earth were almost entirely stripped of volatiles, the era we considered in the preceding chapters.

If we are to appeal to later additions to explain the oceans and the

39

N_2, CO_2, we can also appeal to later additions for trace elements like U, Th, which also may well have been missing in the first place. Even S, and Na, K in the form of alkali feldspar, could well have been late additions to the Earth, along with other less abundant elements having volatile compounds like Hg and Pb. The alkali feldspars condensed at about 1,100 K. According to Table 5.1 such a temperature would place them out by the orbit of Mars. The gravitational field of the Earth, once the Earth had formed itself as a planet, might have clawed such materials back from this more distant region. Or they could have arrived through the Poynting–Robertson effect on small particles. Over time intervals of millions of years there are many ways in which solar system debris could move inwards, from the outer regions to the regions of the Earth and of the other terrestrial planets.

This general point of view implies that, like the oceans and the atmosphere and the limestone, many of the trace elements may well reside only in the surface zone of the Earth. For U, Th, this is known to be true. Heat flow studies of the Earth's crust and upper mantle lead to serious melting problems, causing difficulties in the understanding of the propagation of earthquake waves, unless the heat from the radioactive decays of U, Th, are confined substantially to the top 200 km or so of the Earth's material. How this came about is not at all plausibly explained in the cosmogenic views usually adopted by geophysicists. But if the U, Th, were late additions there is no difficulty at all in understanding how these elements came to be at the surface— they are there because that was just where they landed. But we have still a long way to go toward understanding how the planets themselves came into being before we can discuss questions like this within a detailed framework.

Chapter 8

The Addition of Extra-Solar System Material

After Chapter 1 we have rather lost sight of the interstellar cloud in which the solar nebula formed. No great length of time has yet elapsed however since the solar nebula condensed from the cloud, and the whole proto-solar system must still be embedded in the cloud. If the cloud were of the so-called molecular type, containing the profusion of organic molecules set out in Table 8.1, then its density would typically have been about 10^{-19} g cm^{-3}, low compared to the density of the solar nebula itself, and low compared even to the density in the planetary gases. The planetary gases when they had moved out to the region of the Earth's orbit had a density of about 1.5×10^{-8} g cm^{-3}. The density must have fallen approximately as the inverse cube of the distance from the main solar nebula as the planetary gas continued to move outward, down to about 10^{-12} g cm^{-3} when the regions of Uranus and Neptune were reached, and even this lower density is still millions of times greater than the primordial cloud density. Because of this much greater density one is inclined to think that there cannot be much of a remaining connection between the comparatively diffuse parent molecular cloud and the comparatively dense proto-solar system. But this impression is wrong, as we shall now see.

A spherical cloud of diameter 10^{19} cm and density 10^{-19} g cm^{-3} has a mass of $2.6 \times 10^4 \, M_\odot$. This is formidably large and it would exert a gravitational field on the proto-solar system that would cause it in a time-scale of about 100,000 years to move at a speed of several km s^{-1}

Table 8.1. Molecules detected in interstellar clouds

No. of atoms in molecule							
2	3	4	5	6	7	8	9
H_2	HCN	H_2CO	CH_2NH	CH_3OH	CH_3CHO	$CHOOCH_3$	CH_3CH_2OH
CH	H_2O	HNCO	HC_2CN	NH_2CHO	CH_2CHCN		$(CH_3)_2O$
CH^+	H_2S	H_2CS	CHOOH	CH_3CN	CH_3NH_2		HC_7N
OH	OCS	NH_3	NH_2CN		HC_2CH_3		
CN	HCO	C_3N			HC_5N		
CO	SO_2						
CS	HCO^+						
SiO	HN_2^+						
SiS	C_2H						
NS	HNC						

relative to the gas in the cloud. We have to think therefore of the proto-solar system with its wide-extended ring of planetary gas, at a density of about 10^{-12} g cm^{-3}, moving with a considerable speed through a diffuse gas with density about 10^{-19} g cm^{-3}, and for this relationship to be maintained either for as long as the proto-solar system remains within the cloud or for so long as the ring of planetary gas remains unevaporated from the outer regions of the solar system.

The interstellar gas is much too diffuse to brush away the planetary gas, but by impinging at a speed of several kilometres per second on the planetary gas it can produce a high-temperature boundary zone, which is just what is needed if we are to understand how the hydrogen and helium came to be evaporated from the region of Uranus and Neptune. To evaporate the whole of the hydrogen and helium it is necessary that the total amount of the impinging interstellar gas be of the same order as the mass of the planetary gas. It is easy to estimate the path length through the interstellar gas needed to meet this requirement. Thus the total volume of impinging interstellar gas, with the proto-solar system going through it like an apple-corer, is of order $\pi a^2 l$, where a is the radius of Neptune's orbit, 4.5×10^{14} cm, and l is the path length. The mass of the interstellar gas in this tube-like volume is $10^{-19} \pi a^2 l$, and for this to be equal to the mass of the planetary gas, 1,750 Earth masses (about 10^{31} g), the path length must be about 1.6×10^{20} cm. Because this is almost surely larger than the diameter of the cloud, there is no certainty that the proto-solar system could travel within the cloud for as far as this. But unless some considerable external perturbation of the cloud had the effect of shaking the proto-solar system loose, the likelihood would be that the solar nebula and its **attendant** disk of planetary material would simply oscillate around the interior of the cloud until the hydrogen and helium were evaporated from the exterior of the system. The time-scale for this to happen is given by dividing the path length, 1.6×10^{20} cm, by the speed of the relative motion, and it is of the order of 10 million years.

Turning back to Figure 2.1, and to the discussion in Chapter 2, we saw there that hydrogen and helium will escape more readily if the direction of escape is in the forward sense with respect to the orbital motion about the Sun. But the escaping material then takes away more than the average angular momentum of all the planetary gas, so that the average angular momentum of the remaining gas would decrease steadily as more and more of the hydrogen and helium escaped. This decrease is not compensated by a further addition of angular momentum from the solar nebula, because by now, after 10 million years, the process of angular momentum transfer from the solar nebula to the planetary gas has terminated. The effect of the decrease in average angular momentum would be to cause the ring of planetary gas to shrink, and as the radius of the ring decreased it would become harder to evaporate the remainder of the hydrogen and helium, just because the restraining influence of the Sun became stronger in the manner of Figure 2.1. Inevitably then, a residue of the hydrogen and helium must have remained, for eventually the impinging interstellar gas would not produce a high enough temperature in the boundary zone to evaporate the residue. Rather clearly, the residue of hydrogen and helium was that which went to form the planets Jupiter and Saturn.

It is not difficult to perform a calculation leading to a relation between (i) the ratio of the sum of the masses of Jupiter and Saturn, 413 Earth masses, to the 1,750 Earth masses of the planetary gas before evaporation, and (ii) the ratio of an average radius for the orbits of Uranus and Neptune, say 25 AU, to a suitably weighted average radius for the orbits of Jupiter and Saturn, say 7 AU (closer to the radius of Jupiter's orbit than to Saturn's, because Jupiter is considerably more massive than Saturn). The first ratio is 4.24, and the second is 3.57. What can be deduced by calculation is that the first ratio raised to a power $2(\sqrt{2} - 1)$, which gives 3.31, should be closely the same as the second ratio. The agreement is clearly satisfactorily. In effect, we have succeeded in relating the masses of Jupiter and Saturn to the radii of their orbits.

We come now to the second important aspect of the passage of the proto-solar system through the gas of the cloud. Organic molecules of the kind shown in Table 8.1 would thus be added to our system. Organic molecules that would not have been synthesized in the planetary gases exist copiously in the interstellar clouds. The distinction comes from markedly non-thermodynamic conditions in the clouds, compared to near-thermodynamic conditions in the planetary gas. And since the amount of interstellar gas swept up by the apple-corer process was comparable with the mass of the planetary gas, the amount of the CNO elements added to the solar system in the form of esoteric organic molecules could well have been large. It is true that thermal conditions sufficient to evaporate the planetary hydrogen and helium might have destroyed the more complex molecules, but there must have been times, during the oscillations of the proto-solar system with respect to the cloud, when the speed of the interstellar gas onto the planetary gas was small. In such episodes, or during the initial acceleration of the proto-solar system, the organic molecules would not have been subject to break up, and they could therefore have been added to our system.

It is likely that far more complex molecules than those of Table 8.1 also came to be added similarly to the solar system, a circumstance that may well have been of great consequence for the origin of life in our system. At the end, in Chapter 20, we shall consider the evidence which can be adduced for the widespread occurrence of interstellar polysaccharides. The basic component of polysaccharides is the formaldehyde molecule, H_2CO, whose hydrolysis by the reaction

$$H_2CO + H_2O \rightarrow CO_2 + 2H_2 + energy$$

is the foundation of life. To the polysaccharides we should probably add a whole array of complex prebiotic molecules, the amino acids, nucleotide bases, and sugar-phosphate chains.

Chapter 9

The Aggregation of the Planets

In the previous two chapters we made a big jump in time-scale. Only four thousand years or so was required for the outward transfer of angular momentum from the main solar nebula, whereas 10 million years was required for the evaporation of hydrogen and helium from the periphery of the solar system. Many important things happened in the intervening period, principally because this was the period of the cooling of the solar nebula. Ices, particularly H_2O and CO_2, condensed in the region of Uranus and Neptune. The ices would fall to the equatorial plane of the solar system, forming there a frozen sheet of material. Ice grains would indeed form around previously-existing fine particles composed of the refractory materials that had been brought outward by the gases from the region of the inner planets, and these too would be carried down to the equatorial plane.

There is in Africa, in Gabon, a most curious deposit of uranium ore. The strange thing about this deposit is that the isotope ^{235}U is reduced in its abundance relative to the isotope ^{238}U, to a value which is below the 0.72 per cent possessed by all other known samples of naturally-occurring uranium. It is exceedingly difficult, and likely enough impossible, to understand this difference in terms of any process of diffusive separation of the two uranium isotopes. The suggestion has been made that the ^{235}U abundance was long ago reduced by a naturally-occurring fission cycle, and the deposit in question is sometimes referred to for this reason as the 'Gabon reactor'. Two thousand million years or so

45

ago when this deposit was formed the ^{235}U to ^{238}U ratio was considerably larger than it is at present, making a naturally-occurring reaction a far more feasible possibility than it is at present.* But there is still a serious difficulty with this idea, namely, that so-called neutron poisons, due to other substances in the rocks and dissolved in water, would be likely to prevent a fission cycle from being maintained. I am half-haunted by the thought that fine smoke particles of UO_2, brought down to the equatorial plane of the solar system by water-ice condensing on them, would—if they could manage to get together in a big enough clump—readily form a natural reactor with the ice acting as a moderator. The thought is quite far-fetched, but so are the facts.

Before we finally quit the early first 4,000 years of angular momentum transfer, proceeding next to the much longer time-scale of the aggregation of the planets, it is worth recalling from Chapter 3 the sheet of highly conducting gas that was needed to maintain the operation of the magnetic 'brushes'. We saw that one possible position for this sheet was along the plane of the solar system, in which case the conducting gas would come to contain the solid refractory materials which fell to the plane, since the latter would themselves form a sheet, but a much thinner sheet than that of the conducting gas. Now spurts of highly accelerated charged particles are likely to occur from time to time in the conducting sheet, especially if there were any question of portions of a pulled-out magnetic field biting themselves off in the manner discussed in Chapter 3. The possibility has to be born in mind that such accelerated particles could induce nuclear reactions in the solid refractory materials, reactions that could conceivably lead to observable consequences when investigated by exceedingly delicate modern techniques for the measurement of the isotope ratios of the elements.

After this preamble let us turn to the problem of the aggregation of the planets, concentrating first on the cases of Uranus and Neptune. Our starting point is a sheet of ices, with water-ice the main constituent, and with a total mass of about 2×10^{29} g, the present combined mass of Uranus and Neptune, moving around the newly-condensed Sun. The ices would soon form themselves into considerable chunks, and there would be very many such chunks. To make up 2×10^{29} g needs about 5×10^{22} spherical chunks each 1 m in radius. Too much individuality in the motion of a particular small body leads

---

* For a recent review of this topic, see R. Naudet, *Interdisciplinary Science Reviews*, **1**(1) (1976), 72–84.

to collisions with other bodies. Collisions persist so long as there is any appreciable inconformity in the motions of the various bodies. Eventually the many chunks take up very nearly circular orbits all in the same plane; a thin disk is formed, in a time-scale of only a few orbital revolutions around the Sun, say, about a thousand years.

Although the gravitational force between a metre-sized chunk of water-ice and its nearest neighbours is very weak, the *difference* between the solar gravitational forces acting on two neighbouring chunks is also very small. So the possibility must be considered that, once orderly motion became established, local gravitational effects were able to produce aggregations of small-sized bodies. Taking 2×10^{29} g for the total mass of all the chunks, and taking them to move in orbits with a scale 25 times that of the Earth's orbit, it can be calculated that there would be initial aggregation zones with radii of about 3×10^{10} cm, i.e. about 50 times the radius of the present-day Earth. The total quantity of material contained in such a zone would be about 5×10^{21} g, and the time-scale for the formation of a swarm of bodies of this mass would be only a few orbital periods. After local gravitational forces have pulled such zones of aggregation together, compact icy bodies with radii of about 100 km are formed. Such bodies, in size, mass, and composition would be like large *comets.*

The picture so far is that the initial modest-sized chunks settle quickly into a flat disk, and that this disk then forms into about 40 million much larger bodies of cometary scale. The time required for this first stage is short, not much more than a thousand years. What happens next?

Consider the orbit of a particular comet-sized object. Around very nearly the same orbit there will be many similar objects. The orbits of such objects will never be rigorously the same, however, nor will the orbital periods be rigorously the same. One object will slowly catch up on another. As they thus come close, gravitation can once again cause the objects to join into a still larger body. To specify numbers, the circumference of a circle with radius equal to 25 times that of the Earth's orbit would, if it were divided into bits each of length 6×10^{10} cm (the diameter of our initial zones of aggregation) have about 4×10^4 bits, which is the number of comet-sized objects that can be considered in the present sense to have essentially the same orbit. Even a partial aggregation of so many bodies would evidently yield much larger objects. And as the objects grew in this way, their gravitational spheres of influence also grew—they became able to pull together one to another from still more separated orbits. There is a limit, however, to the efficacy of this gravitational pulling together, which limit can be calculated in the following simple way. Write M for the total mass of the ices, $M = 2 \times 10^{29}$ g, and write m for the mass of the pieces (assumed equal) produced by gravitational aggregation. There will evidently be $N = M/m$ pieces. Each piece can be shown (as we shall see in Chapter 10) to have a gravitational sphere of influence $R(m/2M_\odot)^{1/3}$, where R is the distance of the piece from the Sun, and $M_\odot$ is the mass of the Sun. Now with the ices originally spread from within the orbit of Uranus out to beyond the orbit of Neptune, the N spheres of influence, added systematically from one piece to the next, must cover a range of about 15 AU in R, i.e. about a half of a typical value of R itself, so that

$$NR\left(\frac{m}{2 \, M_\odot}\right)^{1/3} \simeq \tfrac{1}{2}R. \tag{1}$$

Putting $N = M/m$ therefore leads to

$$m = 2\sqrt{2}M\left(\frac{M}{2 \, M_\odot}\right)^{1/2} \tag{2}$$

With $M = 2 \times 10^{29}$ g, $M_\odot = 1.989 \times 10^{33}$ g, (2) gives $m = 4 \times 10^{27}$ g, and $N = M/m$ is about 50.

We have now arrived at the situation shown schematically in Figure 9.1. Many objects with masses like those of the inner planets move in a neat set of concentric orbits with a comparatively small step, $R(m/2 \, M_\odot)^{1/3}$, from one orbit to the next. Thus the step would be about $\tfrac{1}{4}$ of the radius of the Earth's orbit. But of course the well-ordered situation of Figure 9.1 could not persist. With so small a step from one orbit to the next, each planet-like body would be able to exert appreciable gravitational perturbations on its neighbours. The outcome of such perturbations would be to permit a body repeatedly to swop its orbit with a neighbour, working its way repeatedly inward (or outward), so that a particular body would eventually change its position quite appreciably in the general distribution.

To determine by precise calculation exactly what happens in such a complex many-body situation would be difficult, even if the calculation were made with the aid of a fast computer. We do know, however, the outcome of somewhat similar situations for a gas composed of particles each with the ability to deflect the motions of its neighbours. If one were to start the particles of a gas all moving in the same direction, but

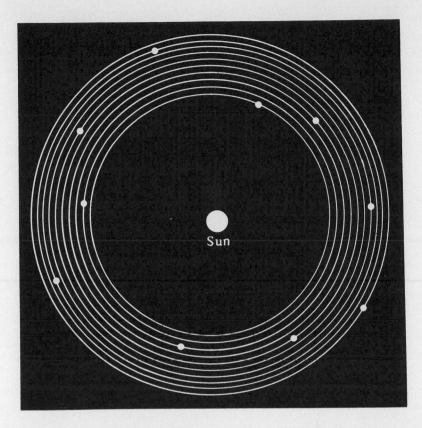

Figure 9.1. A situation during the accumulation of Uranus and Neptune when about 50 bodies of the size of small planets had been formed.

with speeds differing from one particle to another, then the differences of velocity would soon become randomized in their directions. In the situation of Figure 9.1 there are no radial motions and no motions perpendicular to the equatorial plane. Like the gas problem, the effect of gravitational influences between neighbours will be to generate such motions, and they will lead to a far more disordered state of affairs than Figure 9.1, a state of affairs more like Figure 9.2.

In the initial situation of Figure 9.1 there is about a ±10 per cent fluctuation of orbital speed between an object moving in an orbit of radius 25 AU and objects in orbits of radii 20 AU and 30 AU. The effect of gravitational perturbations of one body on another would

Figure 9.2. A random component introduced by gravitational perturbations into the ordered motions of Figure 9.1 leads to this disordered situation.

then be to generate radial motions and motions perpendicular to the
equatorial plane that were of the same order as this ±10 per cent
fluctuation of the initial orbital speeds. The motions perpendicular to
the equatorial plane cause the orbits to depart from the initial plane—
they become inclined to the initial plane at angles up to about 5°. The
volume in which the bodies move is thereby much increased, cutting
down greatly the probability of the bodies joining with each other
through random collisions.

Yet further aggregation depends on actual collisions taking place.
Some collisions would occur already during the disordering process
and they would do so with higher probability than for the fully
disordered situation of Figure 9.2. But collisions occurring during the

disordering process would be unlikely I would suppose to lead to total aggregation, and so, sooner or later, we must come to the problem of calculating the collisions that occur when the motions become fully randomized. I will assume that any collision, even a glancing collision, produces aggregation. A glancing collision would not produce immediate aggregation (and even a head-on collision might not produce complete aggregation) but any contact of two bodies destroys the random relation of their orbits. The orbits become correlated in such a way that both go repeatedly back again to the point of collision, and this greatly increases the probability of their experiencing further collisions, with aggregation eventually taking place.

To calculate explicit numbers, I will suppose that 50 bodies, each with a mass 4×10^{27} g, density 1.5 g cm^{-3}, and radius 8,600 km, move randomly in a washer-shaped region of thickness 2 AU, of inner radius 17.5 AU, and of outer radius 35 AU. The volume of such a region is about 1.9×10^{43} cm^3. Each of the 50 bodies will be taken to sample this volume randomly for the presence of one of the other bodies. Each body does so by drilling out a long curved tube of circular cross-section whose axis is the path followed by the centre of the body. The radius of the cross-section is not just the radius a of the body itself because there will be contact if the radius of a second body is within distance $2a$ of the axis of the tube. And this is without taking account of the gravitational pull of one body on another,* which increases the radius of the tube still more, to about $10a$. Thus if we take a cross-section to have radius 80,000 km we shall not be far wrong. The volume of a length l of the tube is then about $2 \times 10^{20}\, l$ cm^3 (with l in cm). The chance of the body hitting a specified particular member of the ensemble of 50 randomly positioned bodies, while travelling the distance l, is obtained simply by dividing $2 \times 10^{20}\, l$ by the total volume, 1.9×10^{43} cm^3, and the chance of the body hitting any of the ensemble is 50 times greater than this, namely $10^{22}\, l/1.9 \times 10^{43}$. In order for there to be an even chance of such a collision, l must evidently be about 2×10^{21} cm. Now the speed of motion of a body at a distance of order 25 AU from the Sun is about 6 km s^{-1}, and the body thus travels about 2×10^{13} cm year^{-1}. It will therefore travel far enough to encounter some other member of the ensemble of 50 bodies in about 100 million years.

---

* Gravitation increases the effective target radius for collision by a factor 10 above the actual radius because this is about the ratio of the escape velocity from one of the bodies to the encounter velocity between the bodies.

This does not necessarily imply total aggregation into just one or two fully-fledged planets. As coagulation by collision proceeds the number of bodies decreases, and we are then no longer permitted to use multiplication by 50 in a calculation similar to that in the previous paragraph. Indeed, at the last stage of the aggregation it comes down to the chance of one body hitting another specified particular body and there can be no augmentation at all of the ratio of the volume of the swept cylinder to the total volume. But the cylinder now has a larger cross-section. Taking the density within the bodies to stay fixed (which it would do approximately but not completely so because of compressibility) aggregation of 50 bodies into only 2 bodies increases a by $(25)^{1/3}$. And the much more massive 2 bodies also act more strongly gravitationally on each other, causing the tube drilled by either of them to be increased further to about $20a$. Thus the radius of the cross-section is increased above the previous calculation by about $2 \times (25)^{1/3} = 5.85$ and the cross-sectional area is increased by $5.85^2 = 34.2$. This factor largely offsets the former multiplication by 50, leading to much the same estimate, 100 million years, as before. Thus a few times 100 million years, say 300 million years, represents the order of the aggregation times of the planets Uranus and Neptune. Table 9.1 summarizes our results to this point.

A discussion of the aggregation of the inner planets can be given in the same sequence of steps as for Uranus and Neptune. All we need do is to use a total mass of about 10^{28} g instead of 2×10^{29} g, and a mean radius of about 0.85 AU instead of 25 AU. The results are given in Table 9.2.

The time-scales calculated for the final aggregation phases, 3×10^8 years for Uranus and Neptune, and 2×10^6 years for the inner planets, assume that a randomization of the orbits of the colliding bodies continues to be maintained to the end of the aggregation process. This

Table 9.1. Aggregation of Uranus and Neptune

Stage	Mass of bodies (g)	Number of bodies	Time scale (years)
1	5×10^{21}	4×10^7	10^3
2	4×10^{27}	50	$10^6 - 10^7$
3	10^{29}	2	3×10^8

Table 9.2. *Aggregation of the inner planets*

Stage	Mass of bodies (g)	Number of bodies	Time scale (years)
1	5×10^{17}	2×10^{10}	10^2
2	5×10^{25}	200	10^4–10^5
3	5×10^{27}	2	2×10^6
	(Earth, Venus)		
	5×10^{26}	2	
	(Mars, Mercury)		

assumption cannot remain valid, however, when the number of remaining bodies becomes small, because preceding collisions have then had a selective effect, leading to a situation in which the remaining bodies are particularly unlikely to have paths that intersect each other. This anti-correlation effect can greatly lengthen the time required for a final coagulation; fortunately so, since otherwise there would be only one inner planet and not four, and Uranus and Neptune would themselves have coagulated.

What confidence then can we have in the calculated time-scales? It matters greatly in offering an answer to this question how aggregation in stage 3 of Table 9.1 proceeded. We can visualize two very different possibilities, which I will illustrate with reference to Uranus and Neptune. Suppose the 50 bodies at the end of stage 2 (Table 9.1) are all equal, and suppose they join two-by-two to form 25 equal bodies, which again join two-by-two to form 12–13 equal bodies, and so on. Then a stage will come in this egalitarian growth process in which our randomization hypothesis ceases to be maintained. The time-scale for further coagulation will lengthen, because of the anti-correlation effect, and the outcome is likely to be a situation in which many more than two bodies (Uranus and Neptune) still remain after a time equal to the present age of the solar system, about 4.6×10^9 years. But if after stage 2 is completed there are size differences among the 50 bodies, something very different happens. The largest bodies drill out aggregation tubes of greater cross-sectional area than the others, and so they grow more rapidly than the others. If we neglect the gravitational pulling power of a body, the cross-sectional area behaves approximately as $m^{2/3}$, where m is the mass of the body. Gravitation augments this dependence on m, increasing it to at least linearity (cross-sectional area proportional to m), which implies that the bodies grow at least exponentially, like $e^{\alpha t}$ where t is the time and α is a number that is bigger for the largest bodies than it is for the others. Because of this difference of α values, the one or two largest bodies just run away from the smaller ones. And even after the one or two largest bodies have mopped up a half of the available mass, some 25 of the original 50 bodies still remain, and this would seem an amply large enough number for randomization to be maintained amongst them. The time-scales calculated above for stage 3 are thus applicable for the main growth phase of the favoured few big bodies. This second circumstance would seem to be in better correspondence to the situation in the solar system than the egalitarian system. Moreover, it is surely what one would expect to have happened. Egalitarianism is a concept peculiar to the political philosophies of the twentieth century, with but few case examples to support it in either the physical or biological worlds. The biblical saying of 'unto him that hath shall be given' relates more closely to the natural way of things.

In Chapter 8 we saw how some 400 Earth masses of hydrogen and helium could have arrived by inward motion at the region of the planets Jupiter and Saturn. While this explained the positioning of the main constituents of Jupiter and Saturn, did the hydrogen and helium condense by itself into these planets, or did it condense by accretion onto solid nuclei? In the latter case, where did the nuclei come from? The amount of the gaseous hydrogen and helium that went to form Jupiter and Saturn, while apparently substantial at 400 Earth masses, was still too small for there to be much possibility of self-condensation. Accretion onto solid nuclei therefore seems the better explanation. Yet the origin of suitable solid nuclei is not immediately clear. It is true that our calculations have shown the planetary gases, when they first moved outward from the region of the inner planets, to have been carrying an ample supply of refractory particles. But the small amount of fall-out of those particles that occurred in the so-called asteroidal belt between Mars and Jupiter does not encourage the idea that a large fall-out occurred in the region of Jupiter and Saturn. Rather does it seem that little fall-out occurred all the way from Mars to the region of Uranus and Neptune. With an ample amount of ice condensing on them there, fall-out would have occurred, but too far out to be immediately helpful to the problem of Jupiter and Saturn. So I would suppose the solution to the origin of the condensation nuclei of the latter planets to lie in some further consideration.

In the above discussion we took the randomized velocity component of the 50 bodies at the beginning of stage 3 in Table 9.1 to have a fluctuation of ±10 per cent of the speed of motion in a circular orbit. That is to say, any one of the 50 bodies was taken to have a velocity composed of the circular velocity at 25 AU plus a ±10 per cent randomized component. But as with the Boltzmann distribution for the energies of particles in a gas, we may reasonably expect a few of the ensemble of 50 bodies to acquire random velocity components with magnitudes appreciably larger than the average—that is to say, larger than one-tenth of the circular velocity. Thus a member of the ensemble with a 25 per cent departure from the circular velocity at the position of Uranus could follow an orbit which dipped inward to the region of Jupiter, while a body with a 15 per cent departure could follow an orbit which dipped inward to the region of Saturn. Such inward-moving exceptional bodies would encounter resistance from the gas in the region of Jupiter and Saturn, which would cause their orbits to become less elongated, 'rounding' them—but not back to the region of Uranus. They would be rounded up by the drag of the gas into the regions of Jupiter and Saturn themselves, and so could form the condensation nuclei for these planets.

The condition to reach the region of Jupiter involves a more severe departure from the typical fluctuation speed than the condition for a body to reach the region of Saturn—a 25 per cent departure from the circular velocity at the position of Uranus compared to a 15 per cent departure. Thus if one of the 50 bodies in the outer regions managed to dip inward to the region of Jupiter, several may have managed to dip inward to the region of Saturn. So there could have been a competition to form Saturn from a number of condensation nuclei, compared to Jupiter where it may have been hard enough to get just one such nucleus. This detail will assume significance when in the next chapter we come to consider the rotations of the planets.

Chapter 10

The Rotations of Jupiter and Saturn

The problem of the planetary rotations has been strangely ignored by astronomers—not much of a discussion of it can be found in the literature, at any rate in modern times. The basic facts were already contained in Table 2.1. The rotation periods and the axial tilts of Mercury and Venus have been much affected by the friction of the tides which the Sun raises inside the bodies of these planets. Because we therefore have no initial data on the spin of Mercury, nothing is to be deduced from this planet. Something important can be deduced from Venus, however, since the spin of Venus is retrograde. Excluding the curious case of Uranus, all the other planets rotate in a direct sense—that is to say, in the sense of their orbital motions around the Sun. If it were not for the example of Venus, we might easily make the mistake of supposing that the processes which gave rise to the rotations of the planets *forced* their rotations to be direct. Venus shows that, while the processes *favoured* direct rotation, they did not demand it.

The rotation period and the axial tilt of the Earth have also been affected by tidal friction, but more due to the Moon than the Sun. How much of a difference the present-day values of these quantities are from their initial values we cannot say, however, because tidal effects in the remote past cannot be estimated. Tidal effects depend on the configuration of the margin of the land and ocean, and we have little idea of what the configuration was through the first ninety per cent of the Earth's history. An initial terrestrial rotation period of 12 hours and an axial tilt of 15° would be reasonable guesses, but nothing more than that.

I wish to turn in this chapter to the rotation of Jupiter, and at the end of the chapter to that of Saturn, leaving the other planets for consideration in the next chapter. My reason for beginning with Jupiter is that Jupiter is the clearest case, the rotation of Jupiter is the most rapid, it is direct, and the rotation axis is nearly perpendicular to the plane of the orbit. Moreover, it is likely that Jupiter accumulated most of its mass from gaseous hydrogen and helium that moved around the Sun in essentially circular orbits, with the orbital speed at distance R from the Sun given by Kepler's formula, $(GM_\odot/R)^{1/2}$. And to add to these simplifications there may well have been but a single condensation, without collisions occurring among several accumulating bodies to complicate the problem for Jupiter.

Yet even with all these advantages it is still beyond our resources to calculate the accumulation of Jupiter without appeal to further simplifications. But, fortunately, three reasonable approximations change an intractable situation into one that can be handled by quite elementary methods.

The first approximation is to take the elements of gaseous hydrogen and helium to move in orbital planes that are parallel to the plane of Jupiter's orbit. The second is to neglect pressure gradients within the gas in comparison with the effects on the gas of the gravitational fields of Jupiter and of the Sun—this is equivalent to treating the gas as if it were a swarm of independent particles. The third approximation is concerned with the circle shown in Figure 10.1; a circle with its centre at Jupiter, and lying in the plane of Jupiter's orbit around the Sun and with radius $a = R(M_J/2\,M_\odot)^{1/3}$, where M_J is the mass of Jupiter at any stage we wish to consider of its accumulation (here R is the radius of Jupiter's orbit). Outside the circle of Figure 10.1, the gravitational field of the Sun has a stronger controlling influence on particles of gas than has the gravitational field of Jupiter, whereas inside the circle the reverse is the case. And since the transition from one situation to the other is quite sharp (for a circle of twice this radius the field of the Sun would be nearly 10 times more important in its effect on the gas than that of Jupiter, while for a circle of half the radius the reverse would apply) it is a tolerable representation of the problem to suppose that outside the circle the particles move only under the influence of the Sun, while inside the circle they move only under the influence of

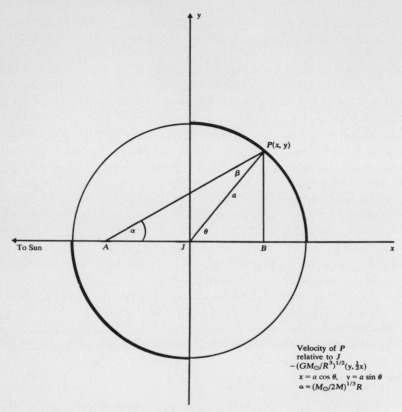

Velocity of *P*
relative to *J*
$-(GM_\odot/R^3)^{1/2}(y, \tfrac{1}{2}x)$
$x = a \cos\theta, \quad y = a \sin\theta$
$a = (M_\odot/2M)^{1/3} R$

Figure 10.1. There is a circle lying in the plane of Jupiter's (J) orbit outside which the Sun has a greater gravitional influence on a moving particle, but inside which Jupiter has the greater influence.

Jupiter. This third approximation dispenses with the need for complex computer calculations.

The critical circle of Figure 10.1 is to be visualized as moving along with Jupiter as the accumulating planet goes around its orbit of radius *R*, the Sun being taken to be away over to the left of the figure. Particles enter the circle from outside over the two heavily marked quadrants, but not over the other two quadrants. Those that enter over the 'north-east' quadrant do so because Jupiter is catching them up in their orbits (being more distant from the Sun the particles move more slowly), while particles enter the 'south-west' quadrant for the opposite reason, because they are catching up on Jupiter. The behaviour of entering particles for one quadrant is entirely the same as for the other quadrant, so that a discussion of either is sufficient. Here I will choose to discuss the 'north-east' quadrant.

It is not hard to show, subject to the simplifications just described, that a particle P entering the circle at a co-ordinate position (x, y) with respect to the axes drawn in Figure 10.1 (axes which move along with Jupiter) will be directed along PA at the moment it enters the circle, where PA is determined by the condition that $\tan \alpha = x/2y$. Nor is it hard to show that the distance AJ is given by $(2y^2 - x^2)/x$, a positive value for this number implying that A lies to the left of J. And the speed of the particle relative to Jupiter is $\sqrt{(GM_\odot/R^3)(y^2 + x^2/4)}$. Thus if the particle is taken to have unit mass its angular momentum about Jupiter is given by

$$\sqrt{\frac{GM_\odot}{R^3} \left(y^2 + \frac{x^2}{4}\right)} \times AJ \sin \alpha = \frac{1}{2} \sqrt{\frac{GM_\odot}{R^3}} (2y^2 - x^2). \qquad (1)$$

The angular momentum is about an axis perpendicular to the plane of Jupiter's orbit, and it is in a direct sense if the particle P happens to have crossed the circle at a point such that $2y^2$ is larger than x^2. Otherwise the angular momentum is retrograde.

Next we notice that for a gas of uniform density, the rate at which particles cross the circle at the position (x, y) is proportional to the product of the velocity and of $\cos \beta$. Thus the rate at which mass crosses the circle at (x, y) is proportional to

$$\sqrt{y^2 + \tfrac{1}{4}x^2} \cos \beta = \sqrt{y^2 + \tfrac{1}{4}x^2} \cos (\theta - \alpha).$$

With a little manipulation this formula can be shown to depend on the angle θ (which can also be used to define the position of P) simply like $\sin \theta \cos \theta$. Thus the rate at which mass crosses the circle is at a maximum at $\theta = 45°$, and the rate falls to zero (rather obviously) both at $\theta = 0$ and $\theta = 90°$.

By the following device we can now determine the *average* angular momentum per unit mass of all the particles crossing the 'north-east' quadrant. The device is to spread a unit of mass over the whole quadrant, giving it a weighting at position θ that goes like $\sin \theta \cos \theta$. In fact, the mass crossing a small bit of the quadrant, the bit between angles $\theta + d\theta$ and θ then turns out to be just $2 \sin \theta \cos \theta \, d\theta$. The average angular momentum is now determined by an integral,* with θ

* $\sqrt{\dfrac{GM_\odot}{R^3}} \displaystyle\int_{\theta_1}^{\theta_2} (2y^2 - x^2) \sin \theta \cos \theta \, d\theta$, with $\theta_1 = 0$, $\theta_2 = \pi/2$.

going from 0 to 90°. The integral can be 'done', as one says (not all integrals can be 'done'!), and the answer is

$$\tfrac{1}{4}R\left(\frac{GM_\odot}{R}\right)^{1/2}\left(\frac{M_J}{2\,M_\odot}\right)^{2/3}. \tag{2}$$

(The last factor in (2) appears when one puts $x = a \cos\theta$, $y = a \sin\theta$, $a = R(M_J/2\,M_\odot)^{1/3}$, into the integral.)

The satisfactory part of our result is that the angular momentum (2) is positive, and is therefore direct. The unsatisfactory part is that, putting $M_J = 1.9 \times 10^{30}$ g, $R = 7.783 \times 10^{13}$ cm, $M_\odot = 1.989 \times 10^{33}$ g, in (2) gives a number that is markedly too big—we have too much angular momentum, a distinct embarrassment of riches. Of course we can argue that Jupiter had smaller mass while it was accumulating, so that the number calculated in this way *should* be too big. Furthermore, we have considered only the situation at the plane of Jupiter's orbit. There will also be additions from parallel planes above and below that of Figure 10.1. The discussion for these will be similar to that given above, but with the critical circle having a smaller radius than $a = R(M_J/2\,M_\odot)^{1/3}$. The outcome for these other planes will be like (2), but with a smaller factor appearing in place of $(M_J/2\,M_\odot)^{2/3}$. The effect of these reductions is insufficient, however. Many years ago, I carried through a calculation like the present one, finding when these other issues were included an average reduction to about $\tfrac{1}{3}$ of (2), say to

$$\tfrac{1}{10}R\left(\frac{GM_\odot}{R}\right)^{1/2}\left(\frac{M_J}{2\,M_\odot}\right)^{2/3}, \tag{3}$$

which is still far too much—it leads to a rotation period for Jupiter of only an hour or so, ten times less than the observed period.

Evidently the calculation has been over-simplified, and it is not at all hard to see where. When we use either (2) or (3) for the angular momentum accumulated by Jupiter, we are implicitly assuming that Jupiter manages to capture all the particles which enter the critical circle across the two heavily marked quadrants of Figure 10.1. Some of the particles may leave the circle through the other two quadrants. To cope with this problem let us give Jupiter an 'effective' radius, r_{eff}, with the property that all particles approaching within distance r_{eff} of the centre of the planet will be captured. Those which pass at greater distances simply escape out of either the south-east or the north-west quadrants of Figure 10.1.

Before we continue the calculation, it is worth noticing that r_{eff} need not be comparable with the normal radius of Jupiter, 7.188×10^9 cm. It can be appreciably larger than this, because a hot atmosphere could well extend outwards to considerable distances from the planet. The particles which Jupiter is accreting have a quite low density (2×10^{30} g distributed in a volume of, say, 2×10^{40} cm^3 has a density of only 10^{-10} g cm^{-3}) and an extension of the atmosphere of Jupiter would readily entangle particles with low density.*

Because we have no clear knowledge of what r_{eff} should be, it is best to set out the problem in such a way that we can see what happens at various values of r_{eff}. It will turn out that there is a critical value of r_{eff} that separates two very different situations.

Going back once more to Figure 10.1, the particle entering the circle at angle θ looks at first as if it is going to pass Jupiter by at a distance $AJ \sin \alpha$. But it actually comes significantly closer because its path is deflected by the gravitational field of Jupiter. It can be shown that the particle comes to a minimum distance r determined by the equation

$$r = AJ \sin \alpha \div \sqrt{1 + \frac{2\,GM_J}{r} \Big/ \left\{ \frac{GM_\odot}{R^3} (y^2 + \tfrac{1}{4}x^2) \right\}}. \qquad (4)$$

Now the numbers turn out to be such that the square-root factor in (4) can be simplified, because in all cases of present interest the unity factor appearing under the square-root sign is unimportant and can be neglected. With this approximation it is easy to show that

$$r = \frac{M_\odot}{2\,M_J R^3} (y^2 + \tfrac{1}{4}x^2)(AJ \sin \alpha)^2. \qquad (5)$$

For capture by Jupiter we therefore require that the right-hand side of (5) shall not be greater than r_{eff}, namely

$$\frac{M_\odot}{2\,M_J R^3} (y^2 + \tfrac{1}{4}x^2)(AJ \sin \alpha)^2 \leqslant r_{\text{eff}}. \qquad (6)$$

Using $AJ = (2y^2 - x^2)/x$, $\tan \alpha = x/2y$, $x = a \cos \theta$, $y = a \sin \theta$, $a = (M_J/2\,M_\odot)^{1/3}R$, an easy algebraic reduction of (6) now leads to the requirement that the magnitude of $3 \sin^2 \theta - 1$, $|3 \sin^2 \theta - 1|$, must satisfy the condition

$$|3 \sin^2 \theta - 1| \leqslant 2^{13/6} \left(\frac{M_\odot}{M_J}\right)^{1/6} \left(\frac{r_{\text{eff}}}{R}\right)^{1/2}. \qquad (7)$$

* The Earth's atmosphere will even entangle high density objects, like rockets and astronauts.

The effect of this condition is to limit the arc of the quadrant from which the entering particles are captured. Particles entering over the rest of the quadrant do not go close enough to Jupiter to be entangled at r_{eff}; they sweep around Jupiter and leave the critical circle of Figure 10.1 through one or other of the lightly marked quadrants. The largest value of θ at which captured particles can enter the circle, θ_2 say, is determined by replacing $|3 \sin^2 \theta_2 - 1|$ by $3 \sin^2 \theta_2 - 1$ and by using the equality condition in (7),

$$3 \sin^2 \theta_2 - 1 = 2^{13/6} \left(\frac{M_\odot}{M_J}\right)^{1/6} \left(\frac{r_{eff}}{R}\right)^{1/2}. \tag{8}$$

For any reasonable choice of r_{eff}, equation (8) always has a solution. And the minimum value of θ at which captured particles enter, θ_1 say, may *either* be zero, $\theta_1 = 0$ *or* θ_1 is determined by

$$1 - 3 \sin^2 \theta_1 = 2^{13/6} \left(\frac{M_\odot}{M_J}\right)^{1/6} \left(\frac{r_{eff}}{R}\right)^{1/2}. \tag{9}$$

The choice between the latter alternatives depends on r_{eff}. If r_{eff} is small enough for the right-hand side of (9) to be less than unity, equation (9) has a solution and θ_1 is this solution. Otherwise $\theta_1 = 0$. The critical value of r_{eff}, given by

$$r_{eff} = 2^{-13/3} R \left(\frac{M_J}{M_\odot}\right)^{1/3}, \tag{10}$$

distinguishes one alternative from the other. For r_{eff} smaller than the right-hand side of (10), equation (9) has a solution and captured particles enter the circle of Figure 10.1 over an arc from θ_1 determined by (9) up to θ_2 determined by (8). For r_{eff} greater than (10), captured particles enter the circle of Figure 10.1 over an arc from $\theta_1 = 0$ up to θ_2 determined by (8).

All we have to do now is to consider an integral like the one in the footnote on page 59, but with the limits of the integral going between the new values of θ_1 and θ_2. The result of evaluating such an integral is

$$\tfrac{1}{4} R \left(\frac{GM_\odot}{R}\right)^{1/2} \left(\frac{M_J}{2 M_\odot}\right)^{2/3} \{3(\sin^2 \theta_1 + \sin^2 \theta_2) - 2\}, \tag{11}$$

which differs from our former result (2) in just the last factor. Now if r_{eff} is less than (10), if θ_1 and θ_2 are determined by equations (8) and (9), subtraction of the latter equations shows that the last factor of (11) is exactly zero. There is *no* average angular momentum, because the arc from θ_1 to θ_2 is adjusted so that the particles adding direct angular momentum are exactly compensated by those adding retrograde angular momentum. However, if r_{eff} is greater than (10) we put $\theta_1 = 0$ and use (8) for θ_2, giving a result which is now direct and *not* zero,

$$\tfrac{1}{4}R\left(\frac{GM_\odot}{R}\right)^{1/2}\left(\frac{M_J}{2\,M_\odot}\right)^{2/3} \times \left[2^{13/6}\left(\frac{M_\odot}{M_J}\right)^{1/6}\left(\frac{r_{\text{eff}}}{R}\right)^{1/2} - 1\right]. \quad (12)$$

For such larger values of r_{eff} *all* the particles giving retrograde angular momentum have been included and yet have been over-compensated by an excess of particles with direct angular momentum.

To calculate the critical value of r_{eff} given by (10), put $R = 7.783 \times 10^{13}$ cm, $M_J = 1.9 \times 10^{30}$ g, $M_\odot = 1.989 \times 10^{33}$ cm. The result is $r_{\text{eff}} = 3.8 \times 10^{11}$ cm, large compared to the present-day radius r_J of Jupiter, 7.188×10^9 cm, but considerably smaller than the radius of the critical circle of Figure 11.1,

$$a = R(M_J/2\,M_\odot)^{1/3} = 6.08 \times 10^{12} \text{ cm}.$$

As an example of a choice of r_{eff} larger than the critical value of 3.8×10^{11} cm given by (10), put $r_{\text{eff}} = 5 \times 10^{11}$ cm in (12). The last factor of (12) then takes the value 0.147, which means that the considerably too large angular momentum previously calculated from (2) is now reduced by a factor of about 7. If one were to allow for all the modifications that changed (2) into (3), and if we also add the final factor of (12), the result for the average angular momentum per unit mass of the particles accreted by Jupiter would be

$$\tfrac{1}{70}R\left(\frac{GM_\odot}{R}\right)^{1/2}\left(\frac{M_J}{2\,M_\odot}\right)^{2/3}. \quad (13)$$

Placed at Jupiter's actual equator, a particle of unit mass with the angular momentum (13) would have a rotational velocity

$$\frac{1}{70}\frac{R}{r_j}\left(\frac{GM_\odot}{R}\right)^{1/2}\left(\frac{M_j}{2\,M_\odot}\right)^{2/3}. \quad (14)$$

Inserting $R = 7.783 \times 10^{13}$ cm, $r_j = 7.188 \times 10^9$ cm, $M_j = 1.9 \times 10^{30}$ g, $M_\odot = 1.989 \times 10^{33}$ g in (14) we get 12.3 km s^{-1}, very close to the actual equatorial rotational velocity of Jupiter.

This last result, while not unsatisfactory, still leaves a lot to be desired. Why $r_{\text{eff}} = 5 \times 10^{11}$ cm? And how did r_{eff} come to be as large as this? To begin the answer to the second of these questions we note a result of a different kind.

The gas density in an isothermal atmosphere of a planet of mass M_p and radius r_p falls off to a non-zero value $\rho_p \exp[-(\mu GM_p/RTr_p)]$ at large distance from the planet. Here ρ_p is the density at the base of the atmosphere, T is the kinetic temperature,* R is the gas constant (equal to 8.317×10^7 when T is in degrees kelvin), and μ is the average mass of the particles of the gas in terms of the hydrogen atom as unit. For a gas composed of hydrogen and helium atoms in a ratio 10/1 (which is about what the planetary gases would have had) μ is about 14/11.

For Jupiter, let us define the 'surface' as a level in the planet where the density ρ_p is $10^{-2} \, \mathrm{g\,cm}^{-3}$. If T is high enough for $10^{-2} \exp[-(\mu GM_j/RTr_j)] \, \mathrm{g\,cm}^{-3}$ to be appreciably greater than the density of the gas crossing the circle of Figure 10.1, then it is reasonable to take r_{eff} to be large, larger even than the value $3.8 \times 10^{11} \, \mathrm{cm}$ given by (10). Now we saw above that the density of the incident planetary gas must have been about $10^{-10} \, \mathrm{g\,cm}^{-3}$, and hence the density $10^{-2} \exp[-(\mu GM_j/RTr_j)] \mathrm{g\,cm}^{-3}$ would appreciably exceed that of the gas crossing the circle of Figure 10.1 provided $\exp-[\mu GM_j/RTr_j]$ was appreciably greater than about 10^{-8}. Thus for

$$\exp-\left[\frac{\mu GM_j}{RTr_j}\right] = 10^{-7} \tag{15}$$

as an example, and with $\mu = 14/11$, $G = 6.67 \times 10^{-8}$, $M_j = 1.9 \times 10^{30} \, \mathrm{g}$, $R = 8.317 \times 10^7$, $r_j = 7.188 \times 10^9 \, \mathrm{cm}$, (15) gives $T = 1.68 \times 10^4 \, \mathrm{K}$.

This condition on the temperature is interesting, because while a gas composed of hydrogen and helium atoms radiates strongly as the kinetic temperature rises above $10^4 \, \mathrm{K}$, such a gas radiates only very poorly at kinetic temperatures below $10^4 \, \mathrm{K}$. Next we notice that the infall of material on to Jupiter generates a speed of about $60 \, \mathrm{km\,s}^{-1}$, which would be sufficient if there were no radiation to drive T for Jupiter's outer atmosphere far above the value $1.68 \times 10^4 \, \mathrm{K}$ obtained in the above calculation. Rapid radiation at high kinetic temperature would soon reduce the temperature, however, down to below $10^4 \, \mathrm{K}$, but not so much below this value. Thus while the above estimate of $1.68 \times 10^4 \, \mathrm{K}$ exceeds the temperature that Jupiter's outer atmosphere

---

* A kinetic temperature is determined by the speeds of motion of the particles. There need be no corresponding thermal field of radiation—nor would there be such a radiation field in the present problem.

could reasonably have had, it is not very much above what it would almost surely have had.

And now at last we are in a position to deduce the rotation period of Jupiter with some precision. The condition (15) gives T essentially proportional to M_J, since r_J is nearly independent of M_J (Saturn with only about one third of the mass of Jupiter nevertheless has a radius that is almost as large as Jupiter, 60,400 km compared with 71,880 km). Hence by reducing M_J to about one-half of its present value we reduce T from 1.68×10^4 K to about 8.4×10^4 K, i.e. to a value that was probably permitted by the poor radiating efficiency of the hydrogen and helium.

Thus up to about half of its present mass, Jupiter would have a large value for r_{eff}, and would therefore acquire direct angular momentum about an axis perpendicular to the plane of its orbit. We do not need to calculate a precise value of r_{eff}. All we need notice is that with r_{eff} rising appreciably above the particular value given by (10), Jupiter would acquire just as much angular momentum as it could store. If need be, the average angular momentum of the added particles could rise to the very large value given by our first calculation, the value given by (3). But the stored angular momentum could not become too large, otherwise Jupiter would spin up to the same kind of rotational instability that we considered for the solar nebula in Chapter 2. It would then shower off the unwanted angular momentum which it could not accommodate. This would happen for a rotation period which we can calculate from an equation

$$\frac{v^2}{r} = 0.54 \frac{GM}{r^2} \tag{16}$$

similar to (2) of Chapter 2, i.e. a period P given by

$$P = \frac{2\pi r}{v} = \frac{2\pi r^{3/2}}{\sqrt{0.54\, GM}}. \tag{17}$$

When we use $M = \frac{1}{2} \times 1.9 \times 10^{30}$ g, $r = 6.5 \times 10^9$ cm (a value for the radius intermediate between the present-day radii of Jupiter and Saturn) equation (17) gives $P = 4.94$ hours.

From $\frac{1}{2} \times 1.9 \times 10^{30}$ g up to the present mass of 1.9×10^{30} g, Jupiter could not maintain a sufficient kinetic temperature in its atmosphere for r_{eff} to be large, and with r_{eff} below the value given by (10), there would be no further addition of angular momentum during the second half of the accumulation of the planet. The average angular momentum per unit mass within Jupiter therefore decreased as $1/M_J$, and with r_J increasing from 6.5×10^9 cm to the present-day radius of 7.188×10^9 cm, the period of rotation increased from the limiting value 4.94 hours to about 12 hours, a result in good agreement with the observed period of 9.8 hours.

When we turn to the planet Saturn we appear at first sight to have something of a mild counter-example. The mass of Saturn, about one third of Jupiter, is low enough for accumulation with large r_{eff}, and so according to the above argument it might seem as if Saturn should be close to the rotational limit of equation (16). For Saturn, putting $M = 5.7 \times 10^{29}$ g, $r = 6.04 \times 10^9$ cm in (17) gives $P = 5.7$ hours, whereas the rotation period of Saturn is actually 10.2 hours.

Saturn is the most rotationally flattened planet in the solar system, however, and is the closest to the stability limit of (16). So our considerations certainly have some correspondence with the facts. And the situation is really better than this comparison of the calculated and observed values of P suggests, because Saturn does not rotate about an axis perpendicular to the plane of its orbit, a clear indication that Saturn accumulated from more than one body of substantial size. The tilt of 29° of the rotation axis implies the existence of a random component in the rotation, due to the collision of two or more such bodies with the following condition satisfied

$$\frac{\text{Random spin component parallel to orbit}}{\text{Sum of random and systematic components perpendicular to orbit}}$$
$$= \tan 29° = 0.55.$$

In adding the random and systematic components in the denominator of the left-hand side of this equation due regard must be had to sense—with the random component being as likely retrograde as direct. The uncertainty here as to sense prevents the use of Saturn as a counter-example. The bodies that went to form Saturn might well have been spinning very close to their stability limits.

Chapter 11

The Rotations of Uranus and Neptune and of the Inner Planets

According to the discussion of Chapter 9, all planets other than Jupiter and Saturn aggregated through a most favoured larger body in a swarm growing through collisions with smaller bodies of the swarm. In this kind of aggregation process each smaller body would be added to the larger body at an impact speed close to the escape speed from the larger body. And as the larger bodies grew towards maturity the escape speeds for the forming planets would be like one or other of those set out in Table 11.1. For whatever planet we have in mind, write V_{esc} for the escape speed and r for the radius.

Now since the chance of a head-on collision is smaller than for a glancing collision (the target area is less) a smaller body adds itself to the larger body with an angular momentum per unit mass that on the average is of order $\frac{1}{2}rV_{esc}$. The factor $\frac{1}{2}$ here makes allowance for the minority of head-on collisions and it also permits us to use the present-day values of r and V_{esc} given in Table 11.1 (during most of the aggregation the product rV_{esc} would be somewhat less than the present-day value). The axis of rotation of the added material is random, differing from one collision to another. So with many collisions of smaller bodies on a larger one there is a great deal of cancelling of the angular momenta, because the sense of rotation in a particular collision may correspond either to a right-handed or to a left-handed twist. The residual angular momentum after N bodies have been added will thus be much less than the $\frac{1}{2}rV_{esc}$ per unit mass added

Table 11.1.

Planet	Radius (km)	Tilt of rotation axis to orbital plane	Escape velocity (km s^{-1})	Equatorial rotational velocity (km s^{-1})	Rotation period	Moment of inertia ÷ Mr^2	Sense of rotation
Mercury	2,439	7°	4.2		58.7 days		Direct
Venus	6,050	174°	10.3		243 days		Retrograde
Earth	6,378	23.5°	11.2	0.465	23^h 56^m	0.333	Direct
Mars	3,394	24°	5.0	0.241	24^h 37^m	0.389	Direct
Jupiter	71,880	3°	61	12.7	9^h 55$^{m\ a}$	0.25	Direct
Saturn	60,400	27°	37	9.9	10^h 38$^{m\ a}$	0.22	Direct
Uranus	23,540	98°	22	3.8	10^h 49^m	0.23	Retrograde
Neptune	24,600	29°	23	2.9	15^h	0.29	Direct

ª For equatorial zone.

in an individual collision. It will in general be less by a factor $\frac{1}{2}\sqrt{N}$, so that the residual angular momentum per unit mass for a planet which grows by adding many bodies in random collisions is of order $rV_{esc}/4\sqrt{N}$. The resulting rotational velocity V_{rot} of the planet is related to this angular momentum per unit mass by the equation

$$k^2 V_{rot} = \frac{1}{4\sqrt{N}} V_{esc}, \qquad (1)$$

where k^2 is the moment of inertia factor given in the seventh column of Table 11.1.

Turning now to the actual planets, we ask:

(i) How do their rotational velocities relate to equation (1)?
(ii) Are their rotational axes randomly distributed?

For Uranus we have $V_{rot} = 3.8$ km s^{-1}, $k^2 = 0.23$, $V_{esc} = 22$ km s^{-1}, so that (1) leads to the deduction that $N = 40$. For Neptune we have $V_{rot} = 2.9$ km s^{-1}, $k^2 = 0.29$, $V_{esc} = 23$ km s^{-1}, and these numbers inserted in (1) lead to $N = 48$. Reference to Table 9.1 shows that in the aggregation of stage 3 we consider Uranus and Neptune to be formed from about 50 smaller bodies. The present results, based on quite different reasoning and on different data, agree very well with our former conclusion.

For Mars we have $V_{rot} = 0.241$ km s^{-1}, $k^2 = 0.389$, $V_{esc} = 5.0$ km s^{-1}, and inserting these values in (1) gives $N = 178$. For the Earth, $V_{rot} = 0.465$ km s^{-1}, $k^2 = 0.333$, $V_{esc} = 11.2$ km s^{-1}, and in this case equation

(1) gives $N = 327$, again in good agreement with the estimate of order 200 in Table 9.2. Neither Venus nor Mercury can suitably be tested in the same way, because their velocities of rotation have been grossly changed by the tidal friction of the Sun.

Unfortunately the second of the above questions cannot be firmly answered, because an issue of randomness cannot be decided by reference to so few examples. But the remarkable case of Uranus, with a spin axis nearly parallel to the plane of the orbit, is indicative of an affirmative answer to it. And so indeed is the retrograde rotation of Venus. On a random basis, the chances of direct and retrograde rotations are equal, and of the six cases available two are indeed retrograde (Uranus as well as Venus).

While the existence of two retrograde rotations amply defends the position against analytical criticism, I must admit to not being wholly satisfied that the rotation axes of the inner planets really were randomly decided. In a random situation there would be an even chance of the rotation axis lying within 30°, not of the pole of a planet's orbit, but of the plane of its orbit. Only for Uranus does this happen. Excluding Mercury and Venus (because of solar tidal friction), and Jupiter and Saturn as being in a quite separate category, we have 3 cases out of 4 with the rotation axis lying within 30° of the pole, not within 30° of the orbital plane. This has the look of a systematic component of rotation being present around the polar axis. I also have the impression of a systematic component that is direct in its sense, a systematic component that happened to be overwhelmed by the random component in the case of Venus, but which has dominated the situations for Mars and the Earth. Can such a systematic component reasonably be contemplated for the inner planets, if not for Neptune?

Turning back to Tables 9.1 and 9.2, the division in time between the stages given there could not be strict. Some larger bodies would reach stage 3 before the aggregation at stage 2 was completed. Stage 2 depends, moreover, on a generally quiescent situation, the smaller bodies not deviating much from circular motion with the Keplerian speed around the Sun. Once several larger bodies had formed there would be perturbation of the quiescent situation, making aggregation in stage 2 go slower and making it more difficult to complete. The effect would be to cause an overlap in time between stages 2 and 3. It is therefore reasonable to suppose that a good deal of smaller debris following roughly circular orbits was still around during the accumulation of Mars and the Earth.

Such debris could have provided the means for direct angular momentum about an axis perpendicular to the orbital planes to have been acquired by the inner planets. The process would need to have been similar to that described in the previous chapter, where in discussing the case of Jupiter we saw that an extended atmosphere (leading to a large value of the quantity r_{eff}) was required. How could such extended atmospheres have been present around the inner planets?

Collisions in stage 3 are between substantial bodies. At the moment of impact, high temperatures would inevitably be generated for a short time, temperatures that were sometimes sufficient to vaporize rock and metal. Not all the impacting material would be vaporized in this way, indeed perhaps only a small fraction of it would be. But even a small fraction could have been sufficient to endow the accumulating planets with temporary gaseous atmospheres. To give a large value of the quantity r_{eff} it would be necessary to have an equation like (15) of Chapter 10,

$$\exp\left[-\frac{\mu GM_p}{RTr_p}\right] = 10^{-7}. \tag{2}$$

Although r_p and M_p, the radius and mass of the planet in question, take values such that the ratio M_p/r_p is much smaller for the inner planets than it was for Jupiter (using present-day values, M_p/r_p is less for the Earth than it is for Jupiter by a factor 28.2) the average weight μ of the particles of gas would be higher for a gas derived from the vaporization of rock and metal than it was for a gas composed mainly of hydrogen atoms. The molecules of vaporized rock would dissociate into atoms, yielding oxygen, which in atomic form has $\mu = 16$. Putting $\mu = 16$ in (2), and $r_p = 6.37 \times 10^8$ cm, $M_p = 5.977 \times 10^{27}$ g, for the Earth, gives $T = 7,480$ K, a temperature not unlike that deduced for the case of Jupiter.

An extended atmosphere would surely occur at the eventual stage of the accumulation to be considered in Chapter 14, when the inner planets came to acquire their volatile materials. Carbon from CO_2 has $\mu = 12$, while hydrogen from H_2O would permit equation (2) to be satisfied for T a little less than 1,000 K (the hydrogen would be molecular at this temperature, with $\mu = 2$).

We may conclude therefore that both the conditions necessary for the Earth and the other inner planets to acquire direct angular momentum by the process discussed in Chapter 10 must have been satisfied on occasion. The question is whether the occasions, inherently temporary in their nature, could have been of sufficient duration for much to have happened. The saving grace here may lie in the very large angular momentum per unit mass that is acquired if r_{eff} is large enough. To end this chapter, suppose material to have been added with r_{eff} large enough for formula (3) of Chapter 10 to be applicable, but with the mass of the Earth replacing M_J in that formula. How much material would need to be added at that angular momentum, viz.

$$\tfrac{1}{10}R\left(\frac{GM_\odot}{R}\right)^{1/2}\left(\frac{M_E}{2\,M_\odot}\right)^{2/3} \text{ per unit mass,} \tag{3}$$

for the whole Earth to have become endowed with direct angular momentum sufficient to give a rotation period of 24 hours? Write m for the mass of the added material, so that the added angular momentum is given by multiplying (3) by m. Since the present-day angular momentum of the Earth is 5.9×10^{40} units, the required value of m is given by

$$\tfrac{1}{10}mR\left(\frac{GM_\odot}{R}\right)^{1/2}\left(\frac{M_E}{2\,M_\odot}\right)^{2/3} = 5.9 \times 10^{40}. \tag{4}$$

Putting $M_E = 5.977 \times 10^{27}$ g, $r_E = 6.378 \times 10^8$ cm, $M_\odot = 1.989 \times 10^{33}$ g, and $R = 1.496 \times 10^{13}$ cm for the radius of the Earth's orbit, equation (4) solves to give $m/M_E = 0.017$. Thus a capture of less than 2 per cent of the Earth's mass would suffice, if it occurred at large enough r_{eff}, to give a significant direct component for the rotation of the Earth. This is so small a quantity of material that perhaps it could have been added during even temporary conditions.

If such a quantity of material, $0.017\,M_E$, were added at a late stage of the Earth's formation with an average density of 3 g cm^{-3} it would form a layer about 65 km deep at the Earth's surface. This result will assume added interest when, in a later chapter, we come to consider the last stages in the accumulation of the Earth. A similar possibility may apply to Mars.

Chapter 12

The Core and Mantle of the Earth

The Earth possesses a core composed mostly of iron. Out to a radius of about 1,250 km the material of the core is solid, but from there to the core boundary at radius $r_c = 3,480$ km the material is liquid. From r_c out to the surface at r_E about 6,370 km, the mantle of the Earth is composed essentially of rock, although some metal deposits, including iron, are to be found close to the surface. Metal is not believed to make much of a contribution to the interior material of the mantle however. Table 12.1 sets out the data on which these conclusions are based. Let us see how this data is to be interpreted.

Solids and liquids have elastic properties—they become squashed if compressed and they rebound when the pressure is released. The density values given in the fourth column of Table 12.1 show the effect of the pressure values in the fifth column. Rock which has a density of about 3.32 g cm^{-3} when it is uncompressed becomes denser as one goes inward to the core boundary, rising to about 5.68 g cm^{-3} at a pressure of some 1.37 M bar (1.37 million bars, the 'bar' being the pressure $10^6 \text{ dyne cm}^{-2}$ generated by a head of 75.006 cm of mercury). Because of the sudden change of the chemical nature of the material which occurs at the core boundary, the density rises abruptly there on crossing inwards, from 5.68 g cm^{-3} to 9.43 g cm^{-3}. And because of the continuing rise of pressure the density increases still further toward the

Table 12.1. The interior of the Earth

Depth (km)	Velocity (km s^{-1}) P	S	Density (g cm^{-2})	Pressure (dyne cm$^{-2} \times 10^{12}$)
33	7.75	4.35	3.32	0.009
100	7.95	4.45	3.38	0.031
200	8.26	4.60	3.47	0.065
300	8.58	4.76	3.55	0.100
400	8.93	4.94	3.63	0.136
413	20° discontinuity			
500	9.66	5.32	3.89	0.174
600	10.24	5.66	4.13	0.213
700	10.67	5.93	4.33	0.256
800	11.01	6.13	4.49	0.30
900	11.25	6.27	4.60	0.35
1,000	11.43	6.36	4.68	0.39
1,200	11.71	6.50	4.80	0.49
1,400	11.99	6.62	4.91	0.58
1,600	12.26	6.73	5.03	0.68
1,800	12.53	6.83	5.13	0.78
2,000	12.79	6.92	5.24	0.88
2,200	13.03	7.02	5.34	0.99
2,400	13.27	7.12	5.44	1.09
2,600	13.50	7.21	5.54	1.20
2,800	13.64	7.30	5.63	1.32
	13.64	7.30	5.68	1.37
2,890	{ Boundary core of			
	8.10		9.43	1.37
3,000	8.22		9.57	1.47
3,200	8.47		9.85	1.67
3,400	8.76		10.11	1.85
3,600	9.04		10.35	2.04
3,800	9.28		10.56	2.22
4,000	9.51		10.76	2.40
4,200	9.70		10.94	2.57
4,400	9.88		11.11	2.73
4,600	10.06		11.27	2.88
4,800	10.26		11.41	3.03
4,982	10.44		11.54	3.17
5,121	9.7		14.20	3.27
5,121	11.16		16.80	3.27
5,200	11.18		16.85	3.32
5,400	11.21		16.96	3.42
5,600	11.25		17.05	3.50
5,800	11.27		17.12	3.56
6,000	11.29		17.16	3.61
6,200	11.30		17.19	3.63
6,371	11.31		17.20	3.64

Earth's centre, but again changing abruptly at the inner core—on this occasion on account of the sudden switch from liquid material to solid.

Because of its elastic properties, waves can propagate through rock. Several modes of propagation are possible, of which the two most important are referred to as *P*- and *S*-waves, the headings of the second and third columns of Table 12.1. The nature of a *P*-wave can be understood by thinking of rock contained inside a cylinder being acted on by a piston. When the piston is under pressure the rock becomes a little compressed. When the piston is released the rock springs back. Repeated oscillations of the piston would produce repeated compressions and expansions of the rock, which is just what happens to rock when a *P*-wave passes through it. During the passage of the *P*-wave the rock is alternately compressing and expanding itself. An *S*-wave, on the other hand, is a twisting form of motion. If a round peg is jammed into a more or less circular hole, one would not seek to loosen it by a continuous turning in either a clockwise or an anti-clockwise sense— this would be only too likely to jam the peg still tighter. What one naturally tries to do is to 'work' the peg loose, and by 'working' the peg we mean putting a series of small oscillatory twists on it, first clockwise, then anti-clockwise, then clockwise again, and so on. When an *S*-wave passes through rock this is the kind of twisting motion which takes place within it. *P*-waves can travel both through liquids and solids, but *S*-waves propagate freely only through solids. *S*-waves are heavily damped in liquids and can at most go through only thin liquid layers.

Earthquakes are generators of both *P*- and *S*-waves, and it is mostly the study of earthquake waves which has yielded the information of Table 12.1. *S*-waves are reflected from the outer core boundary at a radius of 3,480 km, but they do not cross inwards into the core, and it is from this important fact that seismologists inferred that there must be a switch to a liquid material at the core boundary. The comparison of the times at which earthquake waves are received at a number of stations on the surface permits the speeds of the *P*- and *S*-waves at various depths to be determined, and it is these speeds, given in the second and third columns of Table 12.1, that form the basic data for determining the density and pressure values of the fourth and fifth columns. Nowadays it is possible to subject various materials to shocks which generate pressures that are momentarily as large as those in the inner regions of the Earth. The measured speeds of the shock propagation can be compared with the densities generated by the shocks, and so from such experiments the relations of density to pressure and of propagation speeds to density can be obtained for various materials. This laboratory information enables the seismologist to say what kind of material the Earth could be composed of and what it cannot. The first step is to match propagation speeds to the known speeds of earthquake waves at various distances from the centre of the Earth. This gives density and pressure values at various distances from the centre of the Earth (for the particular material, or the combination of materials, under consideration). The pressure and density values are now subject to the condition that the whole Earth must be in mechanical equilibrium—at every level inside the Earth the pressure must exactly balance the weight of the overlying layers. And the density values must also satisfy the requirement that the mass of the whole Earth has to add up to just the mass which the Earth is known (from astronomical studies) to have, 5.977×10^{27} g. These two requirements put very rigorous constraints on what materials it is possible for the Earth to be made of.

A few years ago I wondered (for reasons quite unconnected with the Earth itself) if the core could be made up mostly of copper and zinc instead of iron. After an hour or two of studying the shock-wave experimental data, I saw that it could not. The pressure in copper or zinc needed to obtain propagation speeds as high as the measured speeds of P-waves in the core would blow the Earth violently apart. In a like manner it can be seen from the shock-wave experiments that an old idea held by a minority of astronomers (but never I believe by seismologists or geophysicists) is wrong. The idea was that the core material might be rock which had undergone a phase transition—the experiments do not show a phase transition.

No simple and plausible combination of materials fits both the seismic and the shock-wave data exactly, but a combination of rock for the Earth's mantle and iron for the core comes close to giving a fit. The slight trouble is that, when one requires the Earth to be in mechanical equilibrium and to have its correct mass, the speeds of propagation for iron in the core (as they are obtained from the shock-wave data) come out $0.5 \, \mathrm{km \, s^{-1}}$ or so too low to match the observed speeds of the earthquake waves. So something has to be mixed with the iron. Nickel and iron go closely together in forming condensations within the planetary gases. Some nickel must therefore be mixed with the iron. But nickel is no better than iron. Nickel also gives propagation speeds that are too low, and by about the same margin.* What is needed is a mixture with an element that is lighter, not heavier, than the iron. Mixing with sulphur has been suggested, because in a thermal calculation like the calculations of Grossman, discussed in Chapter 6, iron goes with sulphur to form FeS at temperatures of about 600 K. I doubt, however, that material forming in the planetary gases at a temperature of 600 K could have become incorporated in the deep interior of the Earth. Table 5.1 shows that temperatures of this order belong to the region of Jupiter, from which materials could have arrived at the Earth's surface (as we shall consider in a later chapter) but not at the Earth's centre. Alloys of iron with chromium and titanium would also have the required higher propagation speeds, and since Cr, Ti are also refractory elements (in the sense of condensing at temperatures of order 1500 K) the existence in the Earth's core of a moderate quantity of these elements seems a more reasonable possibility.

Some years ago it was thought that nickel might become segregated from iron in the core, and that having a little higher density (at a given pressure) the nickel sank to the central regions to form the inner core. It was further argued that the inner core is solid because the temperature for melting nickel (again at a given pressure) is somewhat different from the temperature for melting iron. But the melting temperatures are actually so similar, and the unmixing of iron and nickel forming an iron–nickel alloy would be so difficult to understand, that this suggestion too has been abandoned by most geophysicists. Rather does it seem likely that the very central regions are solid because of the relation of the melting temperature to pressure. The temperature required for melting must rise with increasing pressure, so that since the actual temperature must be essentially uniform throughout the core—because metal is a good conductor of heat—one can have the simple situation of Figure 12.1, where the melting temperature curve for iron–nickel alloy rises above the actual temperature just in the very central regions.

While we are not called on to understand how nickel and iron became unmixed from each other, we do have to consider how iron

---

* The properties of iron and nickel are so similar that from a purely seismic point of view the Earth's core could just as well be composed of nickel as of iron. But nickel had less than 10 per cent of the iron abundance in the planetary gases, and for this astrophysical reason the possibility of the Earth's core being mainly nickel instead of iron has never been considered a serious possibility.

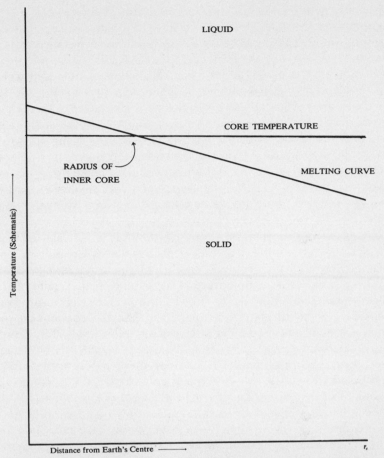

Figure 12.1. A schematic illustration showing that, because of the gradient of the melting curve for the material of the Earth's core, there can be an inner core that is solid.

and rock became unmixed from each other, and this indeed is the primary topic of the present chapter. The Earth's core is clearly different from the mantle, and how did it get that way?

The strongest answer we could give to this question would be to maintain successfully that the iron core of the Earth formed as a complete body before any pieces of rock fell at all upon it, in which case the Earth would have had its iron core from the beginning. This view is hard to maintain, however, even in the present theory, and it is apparently quite impossible to maintain in other theories of the origin of the planets. In other theories, which do not have the segregation of materials implied by the relation of temperature to distance from the Sun that was set out in Table 5.1, the Earth is supposed to have condensed from a nebula of its own. The nebula is supposed to have had all materials—iron and rock-forming elements, water, and other volatiles—present initially as a gas at high temperature. The temperature of the nebula, or protoplanet as it is usually called, is supposed to have declined slowly, with a condensation sequence being set up with respect to temperature, refractories condensing first at temperatures from about 1,200 K to 1,600 K, and volatiles condensing at much lower temperatures. In such a theory the different condensates arise in the same locality, not at markedly different distances from the Sun as in the present theory.

The problem for the iron core of the Earth is clear from the diagrams of Professor Grossman, given already in Chapter 6. These diagrams, especially Figure 6.2, show that the condensation of iron spans much the same temperature range as that of rock minerals, the range from about 1,250 K to 1,475 K. It is true that iron begins condensing at a little higher temperature than the commonest rock-forming minerals, but the difference is small, only about 50 K. This can be seen perhaps more clearly from Figure 12.2 which is also due to Professor Grossman. Figure 12.2 gives the temperatures at which metallic iron, forsterite (Mg_2SiO_4), and enstatite ($MgSiO_3$), condense at various pressures within the gas. Above a pressure of about 10^{-4} bar, iron condenses at higher temperatures, but only by about 50 K, unless the pressure goes higher than one can justifiably take it to have been (10^{-3} bar is usually given as the best estimate). To suppose that the temperature was entirely uniform throughout a protoplanet and that it declined everywhere in exactly the same way is implausible. There would be temperature fluctuations from place to place of at least 50 K (50 K variations in about 1,400 K) and the condensation of rock would inevitably overlap that of iron. So in the protoplanet form of theory the aggregation of solid refractory materials would take place as a heterogeneous mixture of rock and metal, not as a separate core of iron with a mantle of rock accumulating gradually around it. And it then becomes difficult to see how a mixture of a large number of comparatively small pieces of rock and metal once accumulated into the Earth could ever unmix themselves. Big massive lumps of rock and metal would unmix themselves, because the large pressure stresses

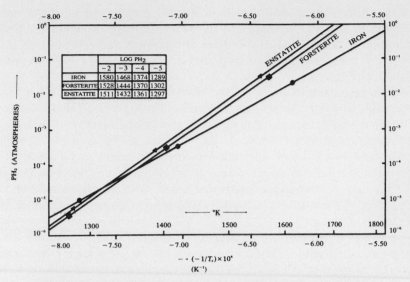

Figure 12.2. The pressure variation of the condensation temperatures of metallic iron and magnesium silicates. Iron condenses before forsterite at pressures greater than 7.1×10^{-5} atm. The lowering of the condensation temperature of enstatite due to the condensation of forsterite is not considered in this diagram [from L. Grossman, 'Condensation in the Primitive Solar Nebula', *Geochimica et Cosmochimica Acta*, **36** (1972), 597].

which big lumps of material of variable density set up would force the higher density lumps of iron to settle to the centre. But the stresses, being much less for small lumps, would be insufficient to cause settling.

The conventional theory, of a cooling gaseous protoplanet leading to the formation of the Earth, fails for a number of other cogent reasons. With an initial chemical composition of the gas of the protoplanet like that of the whole solar nebula (as is usually supposed) the amount of H_2O would exceed the amount of refractories. Yet the mass of the Earth's oceans is only 1.42×10^{24} g, only 0.024 per cent of the total mass of the Earth (5.977×10^{27} g). To the argument that most of the initial supply of water, more than 10^{28} g, evaporated away, one wonders why all of it did not evaporate. Was it a sheer fluke that only 1.42×10^{24} g out of more than 10^{28} g happened to be retained? And how did it come about that the same fluke contrived to happen also for

Mars? And what about the krypton and xenon which would not evaporate, and which would greatly have exceeded the amounts now found on the Earth? Evidently there are so many damaging objections to this conventional point of view that manifestly this was not the way the Earth formed.

Reference back to Table 5.1 shows that, in the theory developed in preceding chapters, a temperature difference of only 50 K still translates into a quite considerable difference in distance from the Sun. The difference of distance between Venus and the Earth is 0.277 AU, which is 4.14×10^{12} cm. The temperature difference of the planetary gas between the two stages when it had moved from the Sun to the distances of Venus and the Earth was 241 K. So if a distance difference of 4.14×10^{12} cm gave rise to a temperature difference of 241 K, a distance difference of $4.12 \times 10^{12} \times (50/241) = 8.55 \times 10^{11}$ cm would give rise to a temperature difference of 50 K, which is by no means a negligible separation. Even so, I do not think the separation to be large enough to prevent iron and rock from becoming far too well-mixed. If all the iron and all the rock condensed at sharply determined temperatures separated by 50 K there might be some chance of forming stage 2 aggregations of iron and of rock separately (for the meaning of stage 2, see Table 9.2). But Figures 6.1 to 6.4 show that condensation takes place over comparatively wide *ranges of temperature*, and that for the most part the temperature range for iron overlaps that for rock. Something more is needed therefore if we are to understand how the Earth came to possess its core of iron.

The distinction I think came about because iron is a good conductor of heat, whereas rock is not. Even a substantial ball of iron would attain a rather uniform temperature, whereas a ball of rock would be hottest at the part of its surface that faced toward the solar nebula. We noted already in Chapter 5 that the radiation incident on a condensate would not be distributed isotropically, as in a purely thermodynamic situation. To the extent that radiation is redirected by the condensates themselves the radiation field would tend more toward an isotropic condition for a large number of small condensed particles than it would for a few bodies of large individual mass. The temperatures of Table 5.1 were calculated for an approximately isotropic situation, whereas the temperature values for an isolated body, with radiation coming only from the solar nebula would be lower by the factor $\sqrt{2}$. This is for a metallic body with good enough heat conductivity for its temperature to be reasonably uniform. For a lump of rock on the other hand, there would be preferential evaporation of material from the hot spot facing the solar nebula. Temperatures of such hot spots would indeed be like those of Table 5.1, higher by $\sqrt{2}$ than for lumps of iron at similar distances from the solar nebula. (No hot spot develops for the iron. The incident heat from the solar nebula is carried through the whole lump, and the whole surface area of the lump then serves to radiate away the incident heat.) A factor $\sqrt{2}$ provides for a marked segregation of iron and rock, since it would fully separate the temperature ranges for condensation (Figures 6.1 to 6.4). Segregation is possible therefore, with the iron condensing nearer to the solar nebula. Indeed a difference less marked than a $\sqrt{2}$ factor would be quite adequate, such as would arise from a radiation field with moderate directivity towards the solar nebula, a radiation field partially on its way to becoming isotropic. Such a half-way situation would arise when there was a moderate degree of re-radiation of heat from the condensed bodies.

In the situation we are now contemplating there would be segregation of iron and rock during stages 1 and 2 (cf. Table 9.2) of the formation of the Earth. The randomizing of orbits occurring in stage 3 must eventually have led to the orbits possessed by the now-large lumps of iron crossing those of the lumps of rock (cf. Figure 9.2). But by now the lumps had attained diameters of the order of 1,000 km, and their masses were at least of the order of 10^{24} g.

There is no doubt that a lump of iron 10^{24} g or more in mass, impacting the accumulating Earth at a speed of order 10 km s^{-1}, would find its way to the Earth's centre. The heat released on impact would liquefy the iron, which would then flow downhill easily with low viscosity. Otherwise stresses in the Earth would need to be sufficient to support imbalances caused by chunks of iron several hundreds of kilometres in diameter, stresses implying horizontal pressure variations of more than 100,000 bar, which is far beyond the yield strength (about 1,000 bar) of solid rock. The present-day liquid condition of iron in the Earth's core is very likely a relic of the impact heating which it received so long ago, for the iron cannot have cooled much, even over the age of the Earth (about 4.6×10^9 years), because of the poor heat conductivity of the overlying rocks of the Earth's mantle.

Rock would also have melted due to impact heating. But the fluid rock would not have plunged into the deep interior where it was shielded from heat loss. The fluid rock remains at the surface, where it cools by radiating heat into space. Given sufficient time, the cooling will lead to solidification, but if impacts occur fast enough the next impact will occur before the rock from the previous impact has had sufficient time to solidify. In Chapter 9 we estimated about 2×10^6 years for the time-scale of stage 3 growth to a final planet. Was this time-scale so short that the Earth's whole rocky mantle was at first in a melted state, or was there sufficient time (on the average) for solidification to occur between one impact of a stage 3 object and the next? To end the present chapter, let us try to answer this question.

Notice first that, given no further impacts, solidification of a molten layer of the Earth's surface must occur from the bottom of the layer upwards. This happens because the temperature difference between the top and bottom of the layer required for the whole layer to be convective, i.e. for it to boil, is less than the temperature difference required to maintain the layer in a liquid condition—the latter being determined by a melting-point curve of the kind shown in Figure 12.1 (but the curve being now for rock instead of iron). The situation is illustrated in Figure 12.3, from which we see that for a given temperature at the surface, a higher temperature is required to maintain fluidity at the base of the layer than is required to maintain a convective outflow of energy from inside the layer to the surface. Hence the surface cannot go solid before the base, because if it did there would be a contradiction. The base still being fluid, the temperature difference from base to surface would be greater than that inferred from the melting curve, and this in turn would be greater than the difference needed to maintain convection, which would immediately carry out a flood of energy to the surface, causing it to melt again.

This means that if the Earth had an initially fluid mantle it had a fluid surface, with a minimum temperature of the order of 2,000 K. Now a body with this surface temperature and with a radius equal to the Earth's mean radius, 6.378×10^8 cm, radiates energy at a rate of 4.64×10^{27} erg s^{-1} (i.e. at a power output of 4.64×10^{17} kW). This is about 1 erg s^{-1} for every gram of rock in the Earth's mantle. So to maintain a fluid condition for as long as 2 million years (about 6×10^{12} s), which was the duration of stage 3 accumulation according to Table 9.2, it would be necessary for each gram of the mantle to have been supplied with about 6×10^{12} erg. This is too much by a large factor. The heat energy possessed by rock at a temperature of order 2,000 K is only about 2×10^{10} erg g^{-1}. So we can be sure that the mantle was *not* in a fluid condition at the end of the accumulation of the Earth. Each major impact of stage 3 must have produced a layer of

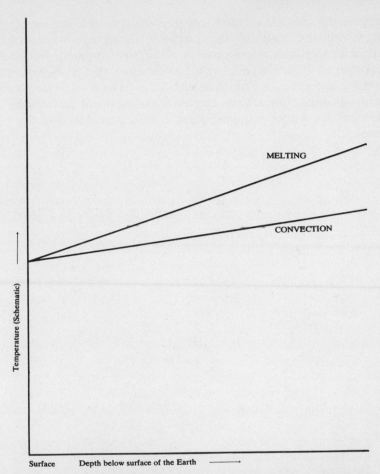

Figure 12.3. If the Earth's upper mantle is liquid it will be convective, because the melting line is steeper than the adiabatic (convection) line. As the liquid cools it will solidify first at the base, because if it attempted as shown here to solidify at the surface, convection would continue, and heat carried upwards would remelt the surface layers.

temporarily fluid rock at the surface. But radiation from the surface solidified the rock, on the average well before the next major impact occurred.

During the temporarily fluid states at the surface, as each rocky body was added, there would be segregation of material, segregation by gravity as heavier materials fell to the bottom of the fluid, and segregation by heat as different minerals solidified at different temperatures. Thus the mantle of the Earth was accumulated in layers, the successive layers corresponding to successive impacts, and with a separation of materials with respect to density and to mineral content occurring separately within each such layer. These ideas will assume added significance in a later chapter when we come to consider the nature of the forces responsible for continental drift and for plate tectonics.

Chapter 13

Impacting Bodies

Bodies impacting at the surface of the growing Earth would do so in the later stages at speeds not much different from the escape speed from the present-day Earth, 11.2 km s^{-1}. Reference back to Table 12.1 shows this speed to be greater than that of seismic waves in the upper part of the mantle, a circumstance which determined that most impacting material would be added to the Earth and not simply sprayed back out into space again.

If one punches hard on a pad mounted on a strong spring the pad moves backward during the punch, and then bounces forward as soon as the punch is over. Looked at from the point of view of the pad, it receives an impulse from outside which it is able to repel through rallying support from the spring. Something of the same sort occurs when a smaller object strikes a large one, as with an object of, say, 100 km radius hitting the surface of the Earth. The immediate surface layer of the Earth forms the pad and the rocks below the surface form the spring. The pad seeks to rally support from below in order to repel the object from outside. This it can only do, however, at the speed of seismic waves, the speed of 'sound' in the rocks, because adequate communication cannot be established from the surface to the underlying rocks at a speed greater than this. Thus if an impacting object strikes at a speed above the velocity of sound in the rock the surface layer cannot rally support fast enough to repel the incoming object. So being unable to lick the invader the surface material joins it. What happens is that the incoming object 'flattens' the surface layer—the outermost layers of rock become compressed. Compression raises the sound speed, enabling the material to communicate more rapidly than before with its surroundings. The most immediately urgent feature of its surroundings is of course the incoming object, which the surface layer now presses more and more strongly against, thereby giving itself a downward motion. This process of pressing against the incoming material goes on until the surface layer finds itself to be moving downward just as fast as the invader. The surface layer has not only joined the invader but has even become its spearhead.

But this only exposes the next layer of the Earth below the first layer to the same problem. The second layer will also become compressed, it will push against the incoming material (which now includes the first layer), and it too will eventually join the downward rush, itself replacing the first layer as the spearhead. And so on to a third layer, and a fourth, and so on—to what? The total momentum of the downward moving cavalcade cannot increase. Indeed the momentum stays much the same as the object picks up more and more of the Earth's surface material. So as the mass of the downward-moving material increases its speed decreases inversely with the mass. Eventually the speed will be so decreased that it falls to the normal sound speed in the outer rocks of the Earth, about 7.5 km s^{-1}. At this stage the layer which the object happens to have reached can rally support from the deeper rocks. The layer which the object has then reached need no longer join the downward-moving cavalcade as a new spearhead. Instead, it feels itself attached to the rocks below. It becomes the pad on the spring which submits to the punch delivered by the cavalcade. The pad retreats for a way in order to coil the spring, and as the pressure in the spring increases the initial object and the picked-up layers are slowed bit-by-bit until the conglomerate comes to a halt. And now at last the Earth has its revenge. The coiled spring releases itself, hurling back the object plus its picked-up layers. The recoil speed is comparable with (but somewhat less than) the speed of sound in the rock, about 7.5 km s^{-1}. This speed is not sufficient in itself to expel the material from the Earth, because it is less than the escape speed of 11.2 km s^{-1}, but it is sufficient to spray material far and wide over much of the surface of the Earth. However, in addition to experiencing a recoil velocity of about 7.5 km s^{-1}, the initial object and picked-up layers may have generated a good deal of internal heat within themselves, which will, if the initial impacting speed was high enough, be sufficient to drive most of the initial object and picked up layers out into space. For a high-enough impacting speed, the Earth will therefore lose material into space. Let us see if we can separate the case where the Earth gains material from the case where it loses material.

Write V_s for the sound speed in the rocks of the Earth's outer mantle, and write V for the impacting speed. The initial energy is thus $\frac{1}{2}V^2$ per unit mass. To slow the speed V down to V_s the incoming object must for each gram of itself pick up $(V/V_s - 1)$ g of the Earth's surface material, so that the mass of downward-moving material is simply increased by the ratio V/V_s. The energy per unit mass has therefore been reduced from $\frac{1}{2}V^2$ to $\frac{1}{2}VV_s$, at the stage where the incoming material has slowed to speed V_s, the stage at which the pad-and-spring provided by the Earth comes into effect. Of this $\frac{1}{2}VV_s$ per unit mass, $\frac{1}{2}V_s^2$ is kinetic energy and $\frac{1}{2}V_s(V - V_s)$ is heat.

Now the pad-and-spring will not be perfectly elastic, so the object and picked-up material will not rebound with speed V_s, but with some lower speed depending on the way in which the kinetic energy $\frac{1}{2}V_s^2$ is shared between the rebounding material and the pad-and-spring. As a rough-and-ready rule I will suppose the sharing to be equal, so that the rebound occurs at speed $V_s/\sqrt{2}$. For materials like solid rock with high V_s, the rebound would probably be more elastic than this, but for softer materials like ice, or rock debris, there would be less elasticity. Thus if we are not to be always chopping and changing the argument the present rule of equal sharing is a fair representation of the average situation, taking all materials into account. So the rebounding material has kinetic energy $\frac{1}{4}V_s^2$ per unit mass, together with heat energy $\frac{1}{2}V_s(V - V_s)$, which it does not share with the pad-and-spring. In total then, the energy per unit mass of the rebounding material is $\frac{1}{2}VV_s - \frac{1}{4}V_s^2$. For the rebounding material to be mostly retained by the Earth, this must be less than $\frac{1}{2}V_{esc}^2$, where $V_{esc} = 11.2$ km s^{-1} is the escape speed from the Earth. For retention we therefore require

$$V < \frac{V_{esc}^2}{V_s} + \frac{1}{2}V_s. \tag{1}$$

Impacting objects in stage 3 of the accumulation of the Earth have incoming speeds given by

$$V^2 = V_{esc}^2 + \left\{ \frac{1}{10} \left(\frac{GM_\odot}{R} \right)^{1/2} \right\}^2, \tag{2}$$

the first term on the right-hand side coming from the fall of the object through the gravitational field of the Earth, and the second coming from a typical encounter between the accumulating Earth and a stage 3 object, for which the encounter speed would be about one-tenth of the orbital speed of the Earth.

Putting $V_{esc} = 11.2$ km s^{-1} and $V_s = 7.5$ km s^{-1} in (1) leads to the condition $V < 20.5$ km s^{-1}, while (2) gives $V = 11.6$ km s^{-1}, little different from V_{esc}. The accumulated Earth would thus have little difficulty in retaining the rebounding material from the impact of a stage 3 object.

For Mars, both V_s and V_{esc} are lower. The speed of sound V_s must be taken at about 6 km s^{-1}, not 7.5 km s^{-1}, because Martian gravity being less than terrestrial gravity, the outer rocks of Mars are by no means as compressed as those of the Earth. Putting V_{esc} equal to 5 km s^{-1} (the appropriate value for Mars) (1) gives $V < 7.2$ km s^{-1}, while (2) gives $V = 5.5$ km s^{-1}. Hence Mars would just about be able to restrain the rebounding material, but the situation is more critical than it was for the Earth.

Venus is a clear-cut situation like the Earth—rebounding material would be retained, but Mercury is still more critical than Mars. The speed of sound in uncompressed iron (iron being the main constituent of Mercury) is about 4 km s^{-1}, lower than in rock, while V_{esc} is 4.2 km s^{-1}, so that (1) gives $V < 6.4$ km s^{-1}, while (2) gives $V = 6.4$ km s^{-1}. The condition for rebounding material to be retained is thus barely satisfied for Mercury. What I think would happen for Mercury is that much rebounding material would initially be lost. But the collision would have the effect of knocking a portion of the material into orbits around the Sun with more of a consonance with that of Mercury itself, giving, for a subsequent encounter, a smaller second term on the right-hand side of (2). Thus if we change the coefficient $\frac{1}{10}$ in this term to $\frac{1}{20}$, (2) gives $V = 4.8$ km s^{-1}, safely below the upper limit of 6.4 km s^{-1} determined by (1).

At the beginning of stage 3 accumulation, before the emergence of any large planet-sized bodies, when the objects were all of a lunar size, the encounter speeds would also be lower than the value $\frac{1}{10}(GM_\odot/R)^{1/2}$ used in the above calculation, because the randomization of the orbits, which determines the encounter speeds, would not yet be fully developed. For collisions of objects with masses of 5×10^{25} g (see Table 9.2) V_{esc} is 1.47 km s^{-1} for rock and 1.71 km s^{-1} for iron. Using $\frac{1}{20}$ instead of $\frac{1}{10}$ as the coefficient in the second term on the right-hand side of (2) and taking $R = 1$ AU gives $V = 2.09$ km s^{-1} for rock and $V = 2.27$ km s^{-1} for iron. These values of the collisional speed are less than V_s for either rock or iron. So there would be no 'pick-up' of material as there was in the case of the Earth. Collisions would be 'softer', with a considerable fraction of the initial kinetic energy going into heat during the encounter. Again using the rule that the kinetic energy of the rebounding material was reduced to one-half of its initial kinetic energy, the rebound speed would be $V/\sqrt{2}$, i.e. 1.48 km s^{-1} for rock and 1.60 km s^{-1} for iron. Thus the rebound speed for rock would be very close to V_{esc}, while that for iron would be just within the value of V_{esc}. The portion of the rebounding material that was retained would evidently be splashed far and wide over the surface of the objects, just as we observe the debris from the collisions of objects that hit the Moon actually to be, as for instance in the bright rays from the crater Tycho.

The value of V_s for ice is less than for rock or iron, about 3 km s^{-1}. Putting $V_{esc} = 20$ km s^{-1} in (2) (for the case of either Uranus or Neptune) gives $V = 20.01$ km s^{-1}, scarcely changed from V_{esc}, while (1) with $V_s = 3$ km s^{-1} gives $V < 135$ km s^{-1}. The condition for impacting material to be retained is therefore satisfied by a very large margin in the cases of the outermost planets.

Chapter 14

The Acquisition of Volatiles by the Inner Planets

The picture we have arrived at for the Earth and Venus is of planets with iron cores and rocky mantles, the rock being a mixture of the various minerals which condensed out of the planetary gases in the manner of Chapter 6. There would also be a differentiation in layers of the mantle with respect to density and mineral content, owing to the temporary melting which occurred for impacting bodies during the stage 3 accumulation of these two planets (Chapter 12). For reasons given in Chapters 6 and 12, Mercury was an accumulation of iron and of some rock, the latter probably being particularly refractory, with corundum a likely possibility. And because iron condensed nearer to the Sun than the rock (Chapter 12 again) we expect Mars to be deficient in iron compared to Venus, Mercury, and the Earth. The initial condition of all four planets was characterized by an almost complete lack of volatile materials, particularly of H_2O, CO_2, NH_3, CH_4, . . . , and of the rare gases. And the surfaces of all four planets were heavily pock-marked with the craters produced by impacts, as we observe the surfaces of the Moon and of Mercury to be even to this day.

By now in our picture the Sun has cooled from its initial primaeval phase of high luminosity. It has become a star with a luminosity about 25 per cent less than the present-day luminosity. Thus the surfaces of the inner planets were dry, heavily scored, and colder than they are at present. The picture is seemingly unpromising as a site for the eventual emergence of life, the scene from an anthropocentric point of view exceedingly dreary.

The drama is not yet over, however, but the scene changes, back again to the region of the outermost planets. Here the aggregation in stage 3 has scarcely commenced. The region is still occupied by a swarm of bodies built from icy crystals which have formed themselves around grains of the refractory materials that were carried out there by the gases, grains that failed to grow large enough to fall out of the gas in the region of the inner planets. We saw in Chapter 9 that the largest bodies to aggregate in stage 2 might have attained masses of the order of 4×10^{27} g, veritable planets in themselves, but although these larger bodies might well have contributed the major fraction of the mass their number would not be anything like as large as the number of smaller bodies, smaller bodies with masses, say, of order 10^{23} g.

As the stage 3 accumulation of Uranus and Neptune got underway, the orbits of the stage 2 aggregations developed the criss-cross pattern of Figure 9.2. This was caused by a randomized velocity component developing from encounters between the bodies. The randomized component required to spread the orbits over the distance range from Uranus to Neptune being about ± 10 per cent of the main orbital component. Exceptionally but inevitably a small fraction of bodies would develop still larger random components of velocity, which would lead to orbits going outside the range from Uranus to Neptune. (Exceptional orbits even among the larger bodies would carry one or two into the region of Jupiter and Saturn, as we already required to be the case in Chapter 10.) And among the numerous swarm of bodies with masses of order 10^{23} g there would be some with orbits that became exceedingly elongated, some that dipped far inwards to the region of the inner planets, rather like the present-day orbit of Halley's comet, shown in Figure 14.1. The condition required to produce an orbit like Figure 14.1 is not a reduction in the speed of motion, but a turning of the direction of motion, a turning which reduces the orbital component of the velocity to about 30 per cent of the velocity $(GM_\odot/R)^{1/2}$ appropriate to a circular orbit. Purely by chance such a situation could arise from the interplay of the gravitational fields of the bodies, particularly in the encounters of smaller bodies with larger ones.

A body which came to follow an orbit like Figure 14.1 would not persist indefinitely in such a path, because further encounters would continue to disturb the motion, and the probability would be for the

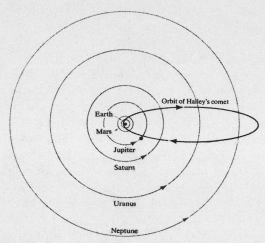

Figure 14.1. Cometary objects with elongated orbits can bring material from the region of the outer planets into the region of the inner planets.

high degree of elongation eventually to be removed. So after a while the body would go back to a more circular orbit, moving once more entirely in the region of Uranus and Neptune. But as one such orbit disappeared from the general assembly of bodies, another would appear.

Suppose one such body dipped each year within the orbit of the Earth, and suppose that it passed into a penny-shaped volume with radius equal to that of the Earth's orbit, 1.496×10^{13} cm, and with thickness 10^{12} cm (i.e. the plane of the orbit of the body not being inclined to the plane of the Earth's orbit by more than about 4°). The area of the curved surface of such a penny-shaped volume is 9.4×10^{25} cm^2, and the chance of hitting the Earth in a random crossing of this area is given by dividing the projected disk of the Earth $\pi \times (6.378 \times 10^8)^2$ cm^2 by 9.4×10^{25} cm^2. Or since the body crosses the curved surface twice, on its way in and on its way out, the chance of collision for each circuit of the body around the Sun is $2\pi \times (6.378 \times 10^8)^2 \div 9.4 \times 10^{25}$, i.e. about 1 part in 30 million.

According to Table 9.1, the accumulation of Uranus and Neptune lasted for 300 million years, and if on the average throughout this period just one body following a path like that of Figure 14.1 came inside the Earth's orbit each year (with a tilt to the plane of the Earth's orbit of not more than about 4°) then the number of collisions with the

Earth occurring in 300 million years would be about 10. And if each colliding body had mass 10^{23} g, and if the Earth managed to retain the colliding material, then the mass acquired by the Earth would be about 10^{24} g. It would consist largely of ices, particularly of H_2O and CO_2, and would be adequate to supply both the oceans and the CO_2 present in the limestone rocks of the Earth.

But would the Earth retain the colliding material? In the preceding chapter we saw that colliding material will be retained provided the value of V given by

$$V^2 = V_{esc}^2 + (\text{Encounter speed})^2 \qquad (1)$$

is less than about 20 km s^{-1}. The encounter speed used in the preceding chapter was for bodies moving wholly in the region of the inner planets, not for orbits of the kind shown in Figure 14.1. For the latter, encounter speeds are higher than the value $\frac{1}{10}(GM_\odot/R)^{1/2}$ used in the preceding chapter. The encounter speed is compounded out of a radial component of relative velocity between the Earth and the body, which can be shown to be close to $\sqrt{2(1-x)} \times (GM_\odot/R)^{1/2}$, and an orbital component of relative motion, which can be shown to be close to $(\sqrt{2x}-1)(GM_\odot/R)^{1/2}$ (provided the orbital motion of the incoming body is in a direct sense around the Sun). Here x is the distance of closest approach of the body to the Sun expressed in terms of the radius $R = 1.496 \times 10^{13}$ cm of the Earth's orbit as unit (i.e. x is in AU). Since these radial and orbital components of relative motion are at right angles, the total encounter speed is

$$[2(1-x)+(\sqrt{2x}-1)^2]^{1/2}\left(\frac{GM_\odot}{R}\right)^{1/2} \qquad (2)$$

The number in the interior of the square brackets now replaces the number $\frac{1}{10}$ appearing in equation (2) of the preceding chapter. Otherwise the discussion of the retention of colliding material is the same as in the preceding chapter. Hence for retention we require the value of V given by

$$V^2 = V_{esc}^2 + [2(1-x)+(\sqrt{2x}-1)^2]\times\left(\frac{GM_\odot}{R}\right) \qquad (3)$$

to be less than about 20 km s^{-1}. Putting $V_{esc} = 11.2$ km s^{-1} for the Earth, and $R = 1.496 \times 10^{13}$ cm, it is easy to see that the condition for retention can be expressed as a condition on the value of x. We must have

$$2(1-x)+(\sqrt{2x}-1)^2 = \frac{V^2 - V_{esc}^2}{GM_\odot/R} \le 0.310. \tag{4}$$

Except for x very close to 1 the first term on the left-hand side of (4) dominates the situation. This term arises from the radial component of the relative motion. For $x = 1$ the left-hand side is 0.172 while for $x = 0.9$ the left-hand side is 0.317. Evidently then, the retention of colliding material demands that the orbit of the incoming body shall not dip much inside the orbit of the Earth. In fact (4) requires that x exceeds about 0.9. Thus for material to be added to the Earth we require the nearest approach to the Sun of a body in an orbit like Figure 14.1 to be less than 1 AU but more than 0.9 AU.

The majority of incoming bodies will almost surely not have values of x between 0.9 and 1 AU, so that usually in collisions the material of the incoming bodies will be sprayed back from the Earth out into space. One might seek to argue that such material would be deflected into new orbits, some of the orbits being perhaps more suitable for the material to be added eventually to the Earth. But this I think to be an insecure argument, since the material, being volatile, would be evaporated into gas and would be subject to being pushed outward away from the region of the inner planets by a wind of particles from the Sun.

In the opposite direction one might seek to argue for a reduction in the amount of volatiles retained by the Earth. If the majority of collisions with bodies in orbits like Figure 14.1 lead to rebound out into space would they not have the effect of also ejecting the volatiles which the Earth apparently managed to acquire from the minority of collisions with x lying between 0.9 and 1? But no, fortunately. Considering H_2O as an example, water acquired in a collision with x between 0.9 and 1 spreads itself in a thin layer widely over the surface of the Earth, and only a modest fraction of the area so covered would be affected by subsequent major collisions. An attempt to argue otherwise, by invoking a widespread rain of much smaller impacts over the whole surface of the Earth, is easily answered by noticing that a new and favourable situation arises for the impacts of smaller bodies.

Once the Earth acquires volatile material (e.g. CO_2) an atmosphere will develop, and once there is an atmosphere small bodies could land 'soft'. Indeed the impulsive pressure which an incoming body experiences on hitting an atmosphere of normal pressure (1 bar) is greater than the yield strength of solid material. The body therefore fragments, and provided it is not too big (not more than about 100 metres diameter) the fragments have time to separate before they reach the solid surface of the Earth. The fragments themselves fragment, so that the incoming object divides itself into a swarm of much smaller pieces each of which is then slowed down by the drag of the atmosphere to terminal speeds that are low enough for the bits to land without further disintegration. This form of 'soft' landing happens today with large meteorites. The landings are in fact so soft that meteorites which are friable enough to be rubbed to powder with the fingers can nevertheless land safely.

The Earth is required to have acquired its first volatiles in at least one hard landing but thereafter it would be possible (if smaller bodies were sufficiently numerous) for a considerable further quantity of volatiles to have reached the Earth in soft landings. Complex organic molecules could have been acquired in soft landings, as (again in fact) organic molecules reach the Earth in meteorites at the present-day.

Returning from small bodies to major impacts a similar, but still more restrictive situation, applies to Venus. Thus for Venus one has a condition similar to (4), with x reinterpreted to mean the nearest approach of the incoming body to the Sun, measured now in terms of the radius of the orbit of Venus as unit, and with 0.239 replacing 0.310 on the right-hand side. The number 0.239 is obtained by inserting $V_{esc} = 10.3$ km s^{-1}, $(GM_\odot/R)^{1/2} = 35.05$ km s^{-1} in the right-hand side of (4), and by again requiring V to be less than 20 km s^{-1}.

Arguing similarly for Mars, with V required to be less than about 6.7 km s^{-1} (if colliding material is to be retained, cf. Chapter 13), and with $V_{esc} = 5.0$ km s^{-1}, $(GM_\odot/R)^{1/2} = 24.14$ km s^{-1}, the left-hand side of (4) would need to be less than 0.0341, a condition which it is impossible to satisfy. (Here x means the distance of closest approach of an incoming body to the Sun measured in terms of the mean radius of the orbit of Mars as unit.) Thus for collision to be possible x must be less than 1, yet already at $x = 1$ the value of the left-hand side of (4) is 0.172, and the left-hand side of (4) only gets bigger still for $x < 1$. Hence Mars cannot have acquired its present-day volatiles, (H_2O, CO_2) from the collisions of elongated orbits like that of Figure 14.1.

There is a second means for acquiring volatiles, however, that is especially favourable for Mars, less so for the Earth, and probably not at all for Venus or Mercury. Figure 9.2 was used to illustrate the criss-crossing that results when a measure of randomness is introduced into the motions of bodies with initially circular orbits (Figure 9.1). What this figure actually shows are the orbits of the so-called short-period comets. All such comets are thought to have once been in orbits that were even more elongated than that of Halley's comet (Figure 14.1). The change from great elongation to comparative circularity has been brought about mostly through the gravitational perturbations of Jupiter. And since the condensation time-scale for Jupiter could not have been as long as the 300 million years that we calculated for Uranus and Neptune, Jupiter would have been able to exert similar influences on the icy bodies which fell from the region of Uranus and Neptune to the inner regions of the solar system in the manner of Figure 14.1. Some of the incoming bodies would have had their orbits rounded up into the situation of Figure 9.2. Once this occurred, any that happened to dip within the orbit of Mars could quite well have been captured by Mars. For now the problem comes back more closely to the situation studied in the preceding chapter, with the impact speed determined by an equation not much different from equation (2) of that chapter. Thus the coefficient involving x in square brackets of the above equation (3) returns to the coefficient used previously—or to a value not too much different from that. Hence there can now be cases in which Mars manages to capture the volatile material of an impacting body which had its origin out in the region of Uranus and Neptune.

Present-day incoming comets begin evaporating their icy material even before reaching the orbit of Mars. Although the evaporation rate is sufficiently slow for much of the solid ice to be able to reach the orbit of the Earth, ice could not persist in solid form at the Earth's distance from the Sun for more than a few months. Present-day comets even at distances from the Sun of about 3 AU gradually lose their ice. We cannot therefore contemplate a long-term situation with icy bodies persisting close to the Earth, and still less so close to Venus or Mercury. For Mars, however, the situation is better, and it would especially be better with the Sun's luminosity lower, as it was by quite a considerable amount for a time-scale of the order of 10 million years, during the epoch before the Sun stabilized itself into an ordinary star. During this period the luminosity was lowered perhaps by as much as a whole stellar magnitude, sufficiently lowered for ices to persist at the orbit of Mars. Thereafter the Sun's luminosity remained lower than it is at present, but by a lesser amount, by about 25 per cent (as we noted earlier), and while this reduction is helpful it may not have been sufficient to give persistence to ices at the orbit of Mars.

The evaporation of ices from bodies rounded up by Jupiter into orbits like those of the short period comets, evaporation which must have occurred except in the above-mentioned period of very low solar luminosity, also leads to interesting consequences. Evaporation releases the refractory grains on which the ices first formed out in the region of Uranus and Neptune. The grains, being small, become subject to the Poynting–Robertson effect, which causes them slowly to lose angular momentum and therefore to spiral gradually inward towards the Sun. The Poynting–Robertson effect arises from the scattering of sunlight by the grains and it leads to a considerable reduction in the angular momentum of a particle, say to a half of its initial value, in a time-scale given by the formula

$$\frac{4\pi R^2 c^2}{L_\odot} \frac{4a\sigma}{3} \tag{5}$$

Here R is the distance of the grain from the Sun, a its radius, σ the density of the material of the grain, c the speed of light, and $L_\odot$ the luminosity of the Sun. Putting R equal to the radius of the orbit of Mars, 2.278×10^{13} cm, $c = 2.998 \times 10^{10}$ cm s^{-1}, $L = 3.8 \times 10^{33}$ erg s^{-1}, $a = 1$ mm and $\sigma = 3.2$ g cm^{-3} for a rocky material, leads to a spiralling time-scale from the orbit of Mars to the orbit of the Earth of about 20 million years. Thus once the refractory nuclei are released from their icy coatings they soon spiral inwards. Moving with the Keplerian speed $(GM_\odot/R)^{1/2}$ in orbits that are nearly circular they are readily subject to capture by Mars, and by the Earth and Moon. Venus may also capture such particles, but only to the extent that they have managed to get by Mars,* the Earth, and the Moon.

In this last process we have at last recovered materials that seemed in the discussions of earlier chapters to have been lost to the outside of the solar system. In particular, we have recovered all the refractory trace elements whose abundances were too low for them to have

---

* From its position, Mars gets the first chance at such spiralling particles. If in the time of exceptionally low solar luminosity there were myriads of fine ice crystals in orbit in the region between Mars and Jupiter, as well as the much more substantial compact bodies that we have considered to this point, the spiralling effect of the Poynting–Robertson process could have led Mars to capture considerable quantities of the volatiles, particularly of H_2O, which is better able to withstand evaporation than CO_2 or NH_3 or N_2.

formed condensates that were large enough in size to fall out of the planetary gases while the gases were still in the region of the inner planets. And perhaps still more interestingly we have recovered the trace elements in a way that puts them at the surfaces of the inner planets. It is relevant to this statement that the inner planets condensed much more quickly than Uranus and Neptune, so that the gradual accretion of small particles continued for long after the main bodies of the inner planets were formed. The process also has an inevitability about it. Inevitably among the very many bodies in the region of Uranus and Neptune some must by chance encounters have had their orbital velocity components reduced sufficiently to bring them at perihelion into the region between Mars and Jupiter. With Jupiter forming more quickly than Uranus and Neptune, it was inevitable that some of these bodies would have had their orbits modified by the influence of Jupiter (and to a lesser extent by Saturn), with the usual effect of producing a rounding of the orbits, just as with the short-period comets. It is then inevitable that the main icy bulk of such bodies (if they are not added by collision to Mars) will eventually evaporate, leaving behind a swarm of refractory particles. Particles with sizes up to a centimetre would then be caused to spiral inward within a time-scale comparable to that required for the aggregation of Uranus and Neptune, 300 million years according to Table 9.1. Conditions for the capture of a reasonable fraction of spiralling small particles, by Mars, the Earth and the Moon, are all good.

How much material might be involved in this way? We calculated in an earlier chapter that of the order of six Earth masses of refractories made their way to the outer regions of the solar system. The amount of material potentially available was therefore large. However, the condition for the development of a highly elongated orbit is unusual—a reduction to about a half in the orbital component of velocity of a body—and only a small fraction of the available material would experience such a situation. And some that experienced it would escape from being seriously influenced by Jupiter, and would eventually return with increased angular momentum back into a more rounded orbit, back into the region of Uranus and Neptune. Furthermore, some that were influenced by Jupiter, some that provided a supply of small particles spiralling towards the Sun would get by Mars and by the Earth, the Moon, Venus, and Mercury, ending their spiralling motion by rejoining the Sun. Yet with all these 'losses', the amounts added at the surfaces of the planets could still have totalled a fraction of a per cent of the six Earth masses carried to the outside of the solar system. The amount to be expected is not calculable in any firm way, unfortunately. But the surfaces of the Earth and Moon are directly available to us, yielding a vast amount of information from which we may hope to decide this question, as I will seek to do in the next three chapters.

Chapter 15

Meteorites and Minor Planets

The solar system divides at a distance from the Sun of order 3 AU, between the inner planets composed largely of refractory materials and the outer planets composed largely of volatiles. In the region between Mars and Jupiter there is very little material, 1.7×10^{24} g has been estimated for this region. A tiny fraction of this small amount reaches the surface of the Earth through the effects of the gravitational fields of the planets, mostly the effects of Jupiter, Mars, and of the Earth itself. The tiny fraction arrives in the form of meteorites, lumps of solid material with masses ranging from a few tens of milligrams up to many tons.

The meteorites are clearly a small fraction of a complex residue of planetary material, which for special reasons happened to fall into orbits with mean radii largely in the range from 2 AU to 4 AU. There were two main ways in which a residue could arrive in this region, one as a minor fall-out of suspended condensates from the planetary gases as the gases moved outward in the manner discussed in earlier chapters. The other way was as a return from the outer regions of Uranus and Neptune, by bodies which happened exceptionally to acquire highly elongated orbits, as in Figure 14.1, and which then became rounded by the effect of Jupiter in the manner of the short-period comets. And within each of these two main routes whereby small quantities of material came to be deposited in the region of the minor planets there were at least two distinct variations. This makes for an intricate situation, a situation I shall attempt to outline in the present chapter.

We saw in Chapter 5 that the deposition of condensates begins with the fall of not very large particles toward the equatorial plane of the solar nebula. Particles with sizes above about a centimetre reach this plane within the available time-scale, which we took to be 1,000 years. The falling of such particles to the equatorial plane is followed by what was described in Chapter 9 as stage 1 condensation, a development in which many particles co-operated through their mutual gravitational fields to produce a number of bodies that were large enough to withstand the viscous drag of the planetary gas. According to Table 9.2, the time-scale for this stage 1 condensation for the region of the inner planets was only about 100 years, less than the time-scale of 1,000 years that it took for the planetary gases to quit the inner region. Stage 1 condensations, which had masses of order 5×10^{17} g, were therefore left behind by the gases as they moved outward.

But not every stage 1 aggregation would be massive enough to withstand the drag of the planetary gases. In a complex many-body situation there would inevitably be some which grew by less than the average amount, to a degree where among very many (according to Table 9.2 there were 2×10^{10} stage 1 bodies) some would be too small to withstand the drag of the outward-moving gas. Such smaller bodies, with diameters up to a few hundred metres, would continue outward for a while with the gases. But not indefinitely, because the gas density decreased with distance R from the Sun, as about the inverse cube of R. And since the drag of the gas would decrease like the density, i.e. proportional to R^{-3}, bodies which had been carried outward from distances of order 1 AU would fall progressively out of the gas as R increased. Some indeed would fall out in this way in the region of the minor planets. Such bodies would be of two main types, there would be lumps of metallic nickel–iron and there would be lumps of rock, the latter with chemical and mineral differentiations of the kind shown by Figures 6.1 to 6.4.

Suspended within the planetary gases were several Earth masses of condensates, in the form of particles that were too small (less than about a centimetre) for them to have reached the plane of the solar system at the stage where the gases were in the region of the inner planets. Some would reach the plane at the later stage when the gases were in the region of the minor planets. This further material would then become available for accretion by the bodies described in the preceding paragraph, as in the stage 2 aggregations discussed in Chapter 9. The masses of the resulting bodies can be estimated if we know the amount of material that was thus made available, If the amount was 10^{24} g and if the stage 2 aggregates had average mass m, then their number N must have been determined by

$$N = 10^{24}/m. \qquad (1)$$

Now as in Chapter 10 the radius of influence of a mass m, in competition with the gravitational field of the solar nebula was $(m/2\,M_\odot)^{1/3}R$, where R was the distance from the Sun. Multiplying this radius of influence by the number of stage 2 bodies gives the spread in distance from the Sun of their orbits, ΔR say, so that

$$\frac{\Delta R}{R} = N\left(\frac{m}{2\,M_\odot}\right)^{1/2} = \frac{10^{24}}{m}\left(\frac{m}{2\,M_\odot}\right)^{1/3}. \qquad (2)$$

Putting $R = 3$ AU, and $\Delta R = 1$ AU, gives m about 10^{20} g, and from (1) N is about 10,000. These results agree very well with the estimated average mass and number of the minor planets. The radius of such a body would typically be about 20 km.

Before returning to the question of meteorites, let us attempt to go to stage 3 of the aggregation process, in which the N gravitational accumulations, at first in almost concentric orbits, as in Figure 9.1, develop a measure of disorder in their motions, as in Figure 9.2. The time required for stage 3 aggregation to complete itself can be estimated by comparing the accumulation of the largest minor planet, Ceres, of mass 6×10^{23} g, with the discussion given in Chapter 9 for the formation of Uranus and Neptune. Because of its low mass, Ceres with a physical radius of 520 km does not gain much of an increase from gravitation in the cross-section of its aggregation-tube. The cross-section is therefore πa^2, with $a = 520$ km, and the volume swept by Ceres in a length l of its path is $8.5 \times 10^{15}\, l$ cm^3 (with l in cm). Taking the volume occupied by the minor planets to be 6×10^{39} cm^3, the chance of Ceres experiencing an encounter over the path length l is given by multiplying the ratio $8.5 \times 10^{15}\, l/6 \times 10^{39}$ by the number of minor planets with randomized orbits. Taking the latter to be 10^4, we require a path length $l = 7 \times 10^{19}$ cm per collision, which Ceres traverses in about 1.2 million years. Over the age of the solar system, 4.6×10^9 years, Ceres would thus be able to acquire about $\frac{1}{3}$ of the ensemble of minor planets to itself. Table 15.1 gives the larger end of the mass distribution of the minor planets. Ceres actually has about $\frac{1}{3}$

Table 15.1. The larger asteroids

Name	Mass (10^{20} g)	Radius (km)	Half of major axis (AU)	Eccentricity of orbit
Ceres	6,000	520	2.77	0.079
Pallas	1,800	280	2.77	0.235
Juno	200	120	2.67	0.256
Vesta	1,000	270	2.36	0.088
Hebe	200	110	2.43	0.203
Iris	150	105	2.38	0.230
Hygiea	600	210	3.15	0.099
Eunomia	400	140	2.64	0.185
Psyche	400	140	2.92	0.135
Davida	300	150	3.18	0.177

of the total mass, about 1.7×10^{24} g, of all the minor planets. Our result is thus in agreement with the facts.

The weakness of this picture is that I have greatly simplified the situation by using the observationally-estimated mass of material within the region of the minor planets, whereas in a fully detailed calculation one would seek to determine what this amount should be. Such a calculation would be much harder than the discussion given above. But weakness or not, the deduced results for the mass of a typical minor planet, and for the accumulation time of Ceres, accord so well with the actual situation that I feel some confidence that the general features of the picture have a satisfactory correspondence with reality. This confidence is very necessary if we are to turn in an equitable frame of mind to the complex problems of the structure and chemical compositions of the various classes of meteorites.

Connection between the theory and the observational and experimental situation for meteorites comes with the distinction in the theory between material added early to the region of the minor planets in the manner of the preceding paragraphs, and material added later from the outer region of Uranus and Neptune. The latter started its history as refractory particles carried by the outward motion of the planetary gases around which ices then condensed, water-ice particularly. With the added mass and size afforded by their icy shells the particles fell to the main plane of the solar system where they became progressively aggregated into bodies of increasing size. Then, in the manner discussed in Chapter 14, a small fraction of these bodies developed elongated orbits as in Figure 14.1.

A few heavy elements, Pb, Bi, In, Tl, remained gaseous as the planetary gases swept in their outward journey through the region of the minor planets. These heavy elements form condensates at temperatures of about 500 K, which was attained on the outward journey in the region between Jupiter and Saturn (Table 5.1). Thus Pb, Bi, In, Tl, would be present in the solid material which formed bodies in the region of Uranus and Neptune, and would hence be present in the material that was eventually added to the region of the minor planets through the rounding of elongated orbits into paths like those of the short-period comets. But the primary distinction between material from the regions of Uranus and Neptune, and that which fell earlier out of the planetary gases on their outward journey, lay not so much in elements like Pb, Bi, In, Tl, as in the contact of the material with water, This in my view forms the essential distinction between the class of meteorites known as carbonaceous chondrites, coming inwards from the region of Uranus and Neptune, and other classes of meteorites, which I think to have been depositions from the gases on their earlier outward journey.

Depending on their mass, icy bodies added to the region of the minor planets may disintegrate, with the eventual evaporation of the ices, leaving small individual particles that become subject to the Poynting–Robertson effect (as we discussed in the preceding chapter). Or if the mass were large enough for the particles to be held together by gravitation, a compact aggregate could be formed. (The difference between the freeing of individual particles and the forming of a compact body is shown today between the meteors, individual small particles freed from comets, and the compact nuclei of comets themselves.)

Mention was made earlier of two variations for the ice-covered material coming from the outer regions. Let us now see how the two variations for this kind of material arise. Ice would condense on particles in the planetary gases, and the ice-coated particles would already fall-out and form themselves into bodies in a time-scale no longer than was required for the luminosity of the solar nebula to decline a little from its high value at radius 2.3×10^{12} cm, the epoch at which the planetary gas first quitted the nebula (Figure 1.1). This cooling time-scale for the first variation was of the order of 10,000 years, short compared with the time-scale required for the evaporation of hydrogen and helium from the periphery of the solar system (by the apple-corer sweeping of gas from the interstellar cloud in which the solar system was embedded, the process we discussed in Chapter 8). The time-scale for the latter process was estimated to be 10 million years. It was also estimated that several further Earth masses of H_2O and of carbonaceous organic material, some of the latter in complex polymerized forms, was added to the exterior of the solar system at this later time. Along with the H_2O and the carbonaceous material there would be the usual complement of very small interstellar solid grains, around which ices would condense as the added gases built to an adequate density for condensation to take place. These further ice-coated solid particles would also fall to the equatorial plane established by the rotation of the solar nebula, aggregating there either into separate bodies or more probably adding themselves at the surfaces of the earlier icy bodies. There would thus be a second type of solid material forming in the region of Uranus and Neptune, a second type distinguished from the first by the circumstances that the solids on which ices condensed *were interstellar in their origin*.

The two types of material would not differ much in their gross chemical compositions. The relative abundances of the elements—a few light gases apart—would be close for both types to the relative abundances within the parent interstellar cloud.[*] The quite different temperature and density histories of the two types of material led on the other hand to important differences in the molecular associations of particular elements.

Interstellar iron would be added initially as small metallic particles, but these would tend to become oxidized at the low temperature in the region of Uranus and Neptune, through the reaction

$$3Fe + 4H_2O \rightarrow Fe_3O_4 + 4H_2.$$

Small particles of metallic iron carried outward by the planetary gas would combine, on the other hand, with hydrogen sulphide to give troilite, through the reaction

$$H_2S + Fe \rightarrow FeS + H_2,$$

[*] It is true that before the planetary gases reached the region of Uranus the solids which fell out of it were chemically segregated, but since the total amount of the solids which reached the outer regions (about six Earth masses) was greater than the amount which was deposited closer to the Sun (about two Earth masses) the fall-out would not produce any great measure of change in the relative abundances of the materials that eventually reached the outer regions.

which would occur as the temperature fell to about 500 K within the planetary gases. Thus iron arrives in the region of Uranus and Neptune in different forms according to whether it comes directly from the interstellar medium or from the outward-moving planetary gases.

Turning now to carbon, the high early temperature of the planetary gases destroyed any complex organic molecules which may have been present in the solar nebula when it first began to condense. Declining temperatures in the planetary gases eventually permitted the existence of complex organic molecules but in the near-thermodynamic conditions within the gases such molecules would not reform themselves. Instead of being in complex organic molecules, carbon would exist only as the simple gas CO until the temperature fell to about 600 K, when CO_2 forms through the reaction

$$CO + H_2O \rightarrow CO_2 + H_2.$$

At about 450 K there is the thermodynamic possibility that methane begins to form through

$$CO + 3H_2 \rightarrow CH_4 + H_2O,$$

but it is to be doubted that much methane could have been formed in this way, since four molecules are required to come simultaneously together for the reaction to happen, a circumstance that in a diffuse gas must be very rare. Sometimes it is possible to arrive indirectly at an end product like $CH_4 + H_2O$ through a sequence of reactions in which other substances play a role. In the present problem it is conceivable that the surfaces of the very many small particles suspended in the gas may also play a part in promoting the intermediate reactions of such a catalytic sequence. Many astronomers and chemists believe this must have been so because hydrocarbons are found in certain types of meteorites. But there are other ways in which hydrocarbons could have been formed, of which the following is an example.

Turning once again to depositions from the planetary gases in the region of Uranus and Neptune, water would condense at a temperature of about 200 K, while solid CO_2 and H_2S would both form at about 150 K. Thus CO_2 and H_2S would often be condensed in juxtaposition with each other, and

$$CO_2 + 4H_2S \rightarrow CH_4 + 2H_2O + 4S$$

occurs with an exothermic energy yield, provided the four resulting sulphur atoms become aggregated into solid free sulphur. Conditions favourable for this reaction might well have occurred when bodies from the region of Uranus and Neptune acquired highly elongated orbits taking them sufficiently inward for heating of a CO_2, H_2S, mix to take place. Subsequent evaporation of both CO_2 and H_2S would then leave free sulphur behind, together with any sufficiently non-volatile hydrocarbons that formed through polymerization from the CH_4.

In contrast to this rather simple situation for the planetary gases, interstellar material could have contributed a rich ensemble of complex organic molecules, including hydrocarbons, that became deposited within the ices that went to form coatings on bodies which had already formed throughout the outer regions of the solar system.

It remains to emphasize the complexity of the processes which led to the origin of the carbonaceous chondrites. The material of these meteorites arrived from the region of Uranus and Neptune through the rounding up of the highly elongated orbits which happened to develop for a small fraction of the many bodies that aggregated in this region (according to Table 10.1 there were 4×10^7 stage 1 bodies). There was a broad division between material of direct interstellar origin and material which arrived in the region of Uranus and Neptune from the outward movement of the planetary gases. Thus planetary-gas material experienced an evolution from high temperatures to low temperatures, whereas interstellar material was added only at low temperature. Because of this difference the organic complexity of the interstellar component was potentially much greater than that of the planetary-gas component. The iron in the interstellar component would be likely to occur as magnetite or as free iron, while the iron in planetary-gas material would occur as troilite. If in the planetary-gas material CH_4 formed through a reaction between CO_2 and H_2S, native sulphur would be produced; and the presence of CO_2 in the planetary-gas material could lead to some carbonates being formed.

In our picture interstellar material condenses on top of planetary-gas material to form bodies with layered structures. Gentle collisions between such bodies could knock off chunks of smaller sizes which could be interstellar only, or planetary-gas only if previous collisions had already pared away the interstellar layer, or indeed complex mixtures of the two types could be generated among the debris from larger collisions. And collisions could of course produce a temporary heating of either kind of material.

Turning now to material deposited in the region of the minor planets on the outward journey of the planetary gases, the ordinary chondrites are identified with this earlier material. Particulate matter having individual sizes up to about 1 cm accumulated on the much larger lumps of iron and rock which happened to fall out of the gases as they moved through the region of the minor planets. The actual chondritic meteorites can be complex mixtures of debris that were knocked off such bodies by collisions. Sometimes the inner cores of the bodies might have become exposed and collisions would then knock off bits of iron or of rock, the rock being essentially similar to the rocks that went to form the inner planets. The bits of iron are identified rather obviously with the iron meteorites.

There is evidence that meteoritic iron formed in large crystals, several metres at least in their dimensions. Such very large crystals have sometimes been interpreted as the product of slow cooling from a melt, but a vapour-to-solid condensation in a gas of slowly declining temperature would be likely also to produce a well-ordered crystallization, because condensation occurs first for the most stable possible form, and this is a single coherent crystal.

The temperatures at which iron meteorites fell out of the planetary gases was about 800 K, the value appropriate for a distance of 3 AU from the solar nebula. It is interesting (and I think significant) that this temperature has an important meaning for the textural properties of these meteorites. Iron changes its crystallographic form as the temperature cools from 800 K to about 650 K. At the higher temperatures the structure is face-centred (taenite) and at the lower temperatures it becomes body-centred (kamacite). The kamacite forms in planes that are parallel to a plane determined by the taenite crystal, with the kamacite layers forming progressively as the material cools. Such a kamacite-layered structure is observed in most iron meteorites and is known as the Widmannstätten pattern.

Cooling through temperatures around 750 K must have been very slow, much slower than the time-scale for the cooling of the solar nebula, which was of the order of 10,000 years. This remarkable fact is inferred from the diffusion of nickel. Iron meteorites have a nickel content which is usually about 10 per cent by mass. Now the nickel fits better into the face-centred form than it does into the body-centred structure. So as the kamacite layers were formed the nickel tended to diffuse back to the parent taenite crystal. But diffusion is a very slow process, and it slows markedly as the temperature falls, dropping essentially to zero at about 650 K. Estimates of the diffusion rates, combined with observation of the amount of diffusion that actually occurred in the kamacite plates of iron meteorites, have led to estimates as long as 10 million years for the cooling time-scale of these objects.

An isolated ball of iron, even a ball some kilometres in diameter, would cool much more rapidly than this, once the luminosity of the solar nebula had declined. However, a ball of iron with an extensive covering of stony material would cool much more slowly, because the heat conductivity of stony material is much less than that of metal. We estimated earlier in the present chapter that the stony coverings acquired by the chunks of iron and of rock (carried from the region of the inner planets and deposited in the region of the minor planets) would grow to radii of the order of 20 km. Sizes of this order would lead to cooling times for an inner core of iron that was adequately long. The cooling time for a ball of hot rock goes as the square of its radius. For a ball of radius about 150 km, the cooling time is of the order of the age of the whole solar system, 4.6×10^9 years. For a ball of radius about 20 km the cooling time would be about 10^8 years.

The present discussion has not taken account of changes which may well have occurred in the meteorites due to subsequent heating. Heating could have occurred in accumulated bodies due to radioactivity, or shock-heating could occur in the collisions of one body with another. The conversion of metallic iron to fayalite ($FeSiO_4$) for example, might have occurred in temperature variations of this kind, and in this respect some of the many problems of the complex mineral content of meteorites might be explained.

Chapter 16

The Chemical Composition of Refractory Materials added to the Surfaces of the Earth and Moon

The Earth and Moon formed in the first place from materials that were highly segregated, without volatiles, and probably without trace elements, except where such elements happened to become associated with common minerals. In Chapter 14 we saw how at a later stage the Earth could have acquired volatile materials, through the impact of bodies from the outer regions of the solar system, bodies in highly elongated orbits like that shown in Figure 14.1. Such impacting bodies would also add materials which in the preceding chapter we associated with the carbonaceous chondrites. The latter contain all the trace elements with abundances that are essentially the same as those of the interstellar gases and hence of the solar nebula itself. In this way the Earth acquired an outer surface-covering of material with non-segregated abundances of all except the lightest and most volatile materials.

The Moon on the other hand could not have retained materials from the impacts it received from bodies in elongated orbits like Figure 14.1. Material in such high speed impacts would rebound from the Moon back into space. But the Moon could have acquired refractory materials with similar relative abundances of the elements. This would happen through the break-up into small particles of similar bodies, brought into the region of the minor planets through the gravitational effect of Jupiter. Small particles, if they should become free of the local gravitational field of the parent body, spiral slowly inward through the

Poynting–Robertson effect (Chapter 14 again). So the Moon would also have acquired trace elements at its surface. Calculations show that the addition per unit surface area to the Moon through the Poynting–Robertson effect is about $\frac{1}{4}$ of the addition per unit surface area to the Earth.

However, the Poynting–Robertson effect brings in other small particles in addition to those with the chemical composition of the carbonaceous chondrites. Indeed, in the last phase of accumulation of the solar system the Poynting–Robertson effect would act as a general sweeping agent for all small particles which happened to be left over from the main processes of aggregation. This can be seen from equation (5) of Chapter 14. Inserting $R = 25$ AU, $\sigma = 3$ g/cm^3, $a = 1$ mm, in this formula gives a sweeping time of about 5×10^9 years, comparable to the age of the solar system.

Much of the swept material would be acquired by the outer planets. After capture by a planet such particles would form an equatorial disk, with the particles spiralling slowly inward towards the planet. If the rate of acquisition of such particles was large enough, giving a captured disk of sufficiently high surface density, subsidiary aggregations would occur, leading to the formation within the disk of a satellite or of several satellites. Indeed, I would suspect most of the satellites of the planets to have originated in this process, with the process still incomplete for the rings of Saturn and of Uranus—incomplete in the sense that the rings in these cases have neither formed into satellites nor has there yet been time enough for their constituent small particles to spiral into the parent planet.

Because of the generality of application of the Poynting–Robertson effect as a sweeping agent, all small particles, whether of highly segregated chemical composition or not, are affected by it. Thus both the Earth and the Moon would receive a general chemical mix-up of material, added mostly over the first few hundred million years of their histories, but still in a reduced measure being added even at the present time.

From these considerations one sees that the chemical nature of the late surface addition to the Moon will be like that added to the Earth in its broad features and this expectation is brought out in a general way by the abundances given in Table 16.1.

Samples of lunar material picked-up from close to the same site show considerable variations, typically by factors of about 3 in the abundances of trace elements, while core samples of lunar soil even

Table 16.1. Abundances of elements in the Earth, Moon, and carbonaceous chondrites[a] (in parts per million unless otherwise stated)

Element	Earth's crust	Moon (average powder unless otherwise stated)	Carbonaceous chondrites (Type I)
Oxygen	46.6%	42.5%	45.3%
Sodium	2.8%	0.28%	0.55%
Magnesium	2.1%	6.37%	9.6%
Aluminium	8.1%	7.2%	0.85%
Silicon	27.7%	21.9%	10.3%
Phosphorus	700	660	1,400
Sulphur	260	500	6.2%
Potassium	2.6%	0.12%	0.05%
Calcium	3.6%	7.6%	1.06%
Scandium	20	32	5
Titanium	0.44%	0.95%	0.04%
Vanadium	140	50	57
Chromium	100	2,900	2,200
Manganese	950	1,570	1,700
Iron	5.0%	11.9%	18.4%
Cobalt	25	22 (fines)	480
Nickel	70	260	1.04%
Bromine	2.5	0.13 (fines)	5
Strontium	390	160	8
Zirconium	170	290	11
Barium	400	220	4
Lanthanum	30	18	0.19
Cerium	60	51	0.63
Neodymium	30	29	0.42
Samarium	6	8.5	0.13
Europium	1	1.2	0.05
Terbium	1	1.85	0.04
Dysprosium	3	6.1 (fines)	0.22
Holmium	1	1.3 (fines)	0.06
Ytterbium	3	3.4 (fines)	0.13
Lutetium	0.5	0.5 (fines)	0.02
Hafnium	3	4 (fines)	0.26
Tantalum	2	0.9	0.02
Iridium	0.001	0.008	0.40
Thorium	7	2.6	0.04

[a] Elements omitted are those for which lunar abundances are uncertain or unavailable.

show considerable variations from layer to layer at the same place. Such intimate differences are what one naturally expects from a conglomerate mode of addition of particles from different parts of the solar system, i.e. different radial distances R, with materials from the different parts being added at different times, and indeed some material being also added as chunks that were not too large to have punctured the surface layer of fine particles. So long as the surface layer was not punctured, impacting objects would experience an unusually low sound speed, about $1 \, km \, s^{-1}$, instead of the $6-7 \, km \, s^{-1}$ that is normal for solid rock. At such low sound speeds impacting objects could land safely with a minimum degree of rebound, as can be seen from the formulae of Chapter 13.

Impacts of objects large enough to reach down to firm rock with much higher sound speeds would scatter debris widely over the surface of the Moon however, and it is rock fragments of this kind that were recovered by the Apollo missions. Impacts at very different places on the Moon's surface can thus lead to debris from different places coming to lie side by side, which no doubt explains the chemical variations that have been found from rock fragments recovered at the same landing site—the variations were not generated *in situ*, they came from variations of material at quite different impact points.

It is a matter of everyday observation that rocks at different sites on the Earth have different chemical compositions. Geologists usually maintain that the enormous variety found on the Earth has arisen from *in situ* processes, notably processes of repeated melting and break-up of the terrestrial surface rocks. However, the Earth would have had great initial variations, just as the Moon has now, and some of the initial differences may well have persisted even through the long history of the Earth. Iron occurs as oxides rather than as fayalite (Fe_2SiO_4), and this probably reflects an initial situation, since iron is found also on the Moon as oxides or to a lesser extent as metal.

Although no lunar sample is like any terrestrial sample in its precise composition, the general situations for lunar and terrestrial surface materials have a number of remarkable similarities. Both have great enhancements of barium, uranium, and thorium. Both have great deficiencies of the platinum metals. Both have excesses of the so-called rare earth elements, excesses and deficiencies being determined relative to silicon (in a comparison with interstellar abundances and with abundances in the Sun).

Potassium and rubidium are greatly enhanced in the surface rocks of the Earth, whereas on the Moon they are not much different from normal. Sodium is moderately enhanced on the Earth and is about normal on the Moon. These differences in the alkali metals represent perhaps the most notable distinction between the outer layers of the Moon and the Earth. Either the Earth had access to some rich source of these metals, a source that was largely denied to the Moon, or the enhancements now found have arisen *in situ* from geochemical processes. The latter 'easy' resolution of the problem is the one most usually taken, but it is noticeable that the condensation of these particular elements occurred in the planetary gases at a temperature of about 1,100 K, which according to Table 5.1 lay in the region of the orbit of Mars. A rather considerable measure of randomization of the orbits of bodies from this region would be needed in order for them to reach the Earth, implying encounter speeds that might well (if the bodies were substantial) have been too high for impacting material on the Moon to have been retained.

Another more fundamental cause of difference between the surface layers of the Earth and Moon could lie in the process of origin of the Moon itself. We have no guarantee that the Moon was always more or less in the position where we find it now. The Earth and Moon could have been bodies which formed in closely similar orbits around the Sun and which became linked even before their main aggregation was completed, although the disparity of their masses, and the Moon's lack of a substantial iron core, argue against this possibility. Alternatively, the Moon could have been captured by the Earth as an almost completely aggregated body, aggregated in a different place and in different circumstances to the Earth.

A very different third possibility is that the Moon was formed from a swarm of small particles which the Earth acquired from the Poynting–Robertson effect, just the same process I would suppose as that which led to the origin of the larger satellites of the outer planets. On this latter basis, one has to understand how a small planet like the Earth acquired enough material to form a satellite comparable in mass to the largest satellites of Jupiter, Saturn, and Neptune, and indeed considerably more massive than any satellite of Uranus. But the Poynting–Robertson effects always work inwards, which gave the Earth access to all small particles from 5 AU inwards (excepting the minority captured by Mars). The area of the plane of the solar system accessible to the Earth was therefore in reasonable proportion to that accessible to Jupiter, for which we may take the range from about 5 AU to 10 AU.

Thus the area accessible to the Earth was about $\frac{1}{3}$ of that accessible to Jupiter, and this is not much different from the ratio of the mass of the Moon to the combined mass of the four main satellites of Jupiter (about $\frac{1}{5}$). Saturn had access to about four times the area available to Jupiter, but its main satellite, Titan, has a mass only about forty per cent of the combined mass of the satellites of Jupiter, suggesting that the small particles available for capture through the Poynting–Robertson effect may have been thinning out in their surface density as the distance from the solar nebula increased to 20 AU. This is consistent with the absence of any large satellite of Uranus, although Neptune has such a large satellite. But Neptune had access to all small particles coming inwards from R greater than 30 AU, so that the potential collecting area was the greatest of all for Neptune.

There is an added attraction to this third possible mode of origin of the Moon, for it may explain why Venus has no satellite. The comparable collecting area for Venus extends only from 0.72 AU to 1.00 AU, giving access to a much smaller area, about $\frac{1}{50}$ of that available to the Earth. Venus would certainly have acquired some particles, which eventually spiralled into the planet itself instead of forming a satellite, presumably because the density of the acquired particles as they moved around Venus in a disk never rose high enough for their self-gravitation to become sufficient to cause them to aggregate into a satellite.

An interesting fact may be noted which argues for the very last additions to the Moon to have taken place in the vicinity of the Earth. Provided the Moon was in much its present position the rate of addition per unit area of small particles to its surface must have been about $\frac{1}{4}$ of that to the Earth. Thus if uranium and thorium were added late to both the Earth and Moon the rate of heat generation from radioactivity at the present-day should be four times greater per unit surface area for the Earth than it is for the Moon, a deduction that is correct. Thus the present-day heat flow, believed to be derived from radioactivity, is 69 erg cm^{-2} s^{-1} for the Earth's surface, while the mean value for the Moon is about 16 erg cm^{-2} s^{-1}. The heat flow does not vary much over the surface of the Earth (hot spots close to molten rock apart) whereas variations from 12 ergs cm^{-2} s^{-1} up to 21 ergs cm^{-2} s^{-1} have been measured for the Moon. Such variations suggest that initial irregularities in the disposition of radioactive materials still persist on the Moon, whereas on the Earth such irregularities have been smoothed away by horizontal movements of the surface material.

The rocks picked-up at landing sites on the lunar maria are of basaltic composition with ages less than 4×10^9 years. Now the addition of small particles must have continued for a time-scale of at least the accumulation time for Uranus and Neptune, which was given in Table 9.1 as 3×10^8 years. Thus the addition of material continued for a time scale of the order of the difference between the age of the solar system itself and the measured ages of most lunar rocks. The latter ages represent the times for which the rocks have been chemically-closed systems, with material neither being added nor taken away from them, particularly the radioactive elements and their products on which the age determinations are based (nowadays usually strontium and rubidium). Melting destroys the chemical closure of particular samples, so that age determinations are to be interpreted as the time elapsed since the last melting occurred, for lunar rocks mostly about 3.8×10^9 years ago.

There are two ways of interpreting these lunar age determinations. The interpretation favoured by nearly all lunar geologists is that extensive melting of the Moon occurred about 3.8×10^9 years ago, due to the heat released by radioactive potassium and by the isotope ^{235}U of uranium. The second possibility is that the last appreciable impact melting of the surface material of the Moon occurred about 3.8×10^9 years ago. In support of the first interpretation one can argue that the difference $4.6 \times 10^9 - 3.8 \times 10^9 = 8 \times 10^8$ years is roughly comparable with the half-lives of ^{40}K and ^{235}U, so that a considerable fraction of the radioactive heating from these nuclei had been released by the time of the melting of the material (i.e. the material of the rocks picked-up at the lunar surface). In support of the second interpretation, the time difference of 800 million years, while greater than the time estimated for the main aggregation of Uranus and Neptune, is probably a reasonable estimate for the completion of that process. During this interval bodies would be continuously directed into elongated orbits like that of Figure 14.1. thereby permitting material to be added at the surfaces of the Moon and Earth. It is not of course necessary that impact melting should be the only melting experienced by lunar rocks. The rocks could have been melted when their constituent material was first added to the Moon, or even at a still earlier stage when the material first condensed within the planetary gases, when indeed its rock type (e.g. basalt) may first have been established.

A surface layer with a low speed of sound necessarily implies much melting with consequent geochemical segregations of material occurring in recrystallizations from the resulting melts. For an impact speed V of 3.5 km s^{-1} onto the Moon, and for a sound speed $V_s = 1 \text{ km s}^{-1}$, the 'pick-up' phenomenon of Chapter 14 occurs, with the impacting mass acquiring material in its path until there is a slowing down to speed V_s. At this stage the kinetic energy per unit mass is $\frac{1}{2}V_s^2$ and the heat energy is $\frac{1}{2}V_s(V - V_s)$ per unit mass, i.e. $1.25 \times 10^{10} \text{ erg g}$, sufficient to raise the temperature of the impacting material plus pick-up material to the melting point of rock. Thus impact heating due to infalling objects must produce pools of molten material leading to the fusing of small particles and to wide variations of chemical composition on recrystallization. Materials not fitting readily into common minerals can become highly concentrated in this process. I would attribute the remarkable variations of K content shown by samples of lunar material to be due to this effect. All other non-miscible substances would show comparable effects and would be correlated with the potassium abundance.

Chapter 17

The Effect of Radioactivity on the Earth's Outer Mantle and Crust

We have seen in Chapter 12 that the mantle of the Earth formed in layers, caused by the addition of impacting objects of considerable size, probably indeed some as large as the Moon. Each such layer was initially in a fluid state because of impact heating. Solidification soon took place, however, from the bottom upward, with a temperature gradient essentially that of the curve of melting-point versus pressure. We also saw that melting would sort each fluid layer with respect to density, with the heavier materials falling to the bottom of the layer. As the Earth accumulated with layer after layer being added on top of each other, the pressure within the lower lying layers would rise, taking the temperature required for melting well above the actual temperature. Thus while the material of the outer layers would have temperatures close to their melting temperatures, lower layers would be much below melting temperatures. Subsequent heating processes (due to radioactivity) would thus be inherently more likely to melt the upper rather than the lower layers; that is to say, the first few hundred kilometres below the surface would be susceptible to melting, whereas the material at greater depths would require much more heat if it were to melt.

The main four nuclei responsible for long-term release of heat through radioactivity are ^{40}K, ^{232}Th, ^{235}U, and ^{238}U, with half-lives of 1.28×10^9 years, 1.41×10^{10} years, 7.1×10^8 years, and 4.51×10^9 years respectively. Now of all materials I would expect Th, U, to be added last to the Earth. The abundances of these elements, the first to condense as the planetary gases quitted the solar nebula, were so low that only very small grains could have formed, grains that would have been easily carried by the planetary gases to the outer regions of the solar system. Their return to the region of the inner planets was quite likely through the Poynting–Robertson effect, with lifetimes for inward spiralling that were much longer than the time-scale for the accumulation of the main body of the Earth.

Even potassium could not have been added to the Earth during the main phase of its aggregation. The alkali feldspars condensed (according to the temperatures of Table 5.1) at distances from the solar nebula well beyond the radius of the Earth's orbit, and it would only be with a considerable randomization in the motions of condensations containing potassium-bearing minerals that they would become subject to capture by the Earth. The very high potassium content of the surface rocks of the Earth (averaging about 2–3 per cent by mass) points in this direction, although the potassium content of the surface rocks has almost surely been increased by later igneous processes.

Suppose for the purpose of making a calculation of heat production from ^{40}K that all potassium was contained initially in the so-called asthenosphere with a depth of about 300 km. Suppose further that owing to subsequent segregation processes most of the potassium has come to be concentrated in a surface layer of thickness 10 km, where it is sufficient to give a 3 per cent concentration by mass to that layer. Then the initial concentration by mass throughout the asthenosphere must have been about 1,000 parts per million (ppm), comparable to the potassium content of lunar soil. Now the present-day ratio of ^{40}K to total potassium is 1.18×10^{-4}, while the content of ^{40}K when the potassium was first added to the Earth was about 12 times higher than this, about 1.4×10^{-3} of total potassium. So a total potassium content of 1,000 ppm for the asthenosphere would imply an initial ^{40}K content of about 1.4 ppm.

Most of the ^{40}K has by now decayed, with energy release of about $1.7 \times 10^{16} \, \text{erg g}^{-1}$ of ^{40}K. Hence the heat released per gram of rock in the asthenosphere would be about $1.4 \times 10^{-6} \times 1.7 \times 10^{16} \, \text{erg} = 2.4 \times 10^{10} \, \text{erg}$, sufficient (if the heat remained trapped) to raise the temperature of the rock by $2.4 \times 10^{10}/C_p \, \text{K}$, where C_p is the specific heat of the rock. Since C_p is about $10^7 \, \text{erg g}^{-1} \, (\text{K})^{-1}$, the temperature rise would be about 2,500 K. The rocks would melt for so great a heating, convection would ensue, and the above assumption, that the heat remained trapped, would be vitiated.

No account has been taken in this calculation of heat production from Th, U, because these elements would initially be so close to the surface that conduction could very well carry their radioactive heat production to the actual surface, whence it would be radiated into space. However, if the energy production from ^{40}K is able to heat the asthenosphere sufficiently for melting to occur, convection must arise, leading to the possibility of downward transport of a portion of the Th, U, causing the heat production from these elements to become distributed at considerable depths, possibly depths of 100 km or more.

The present-day heat production from Th, U, occurring in the outer layers of the Earth, is known to be about 10^{28} erg year^{-1}. The average heat production taken over the age 4.6×10^9 years of the Earth has been greater than this, by a factor which I estimate to be about 3.7, giving a total energy production of $3.7 \times 10^{28} \times 4.6 \times 10^9 = 1.7 \times 10^{38}$ erg. Now taking the mean density of rock as 3.2 g cm^{-3}, the total mass in the asthenosphere is about 4.6×10^{26} g, so that the heat production per gram of rock from Th, U, averaged for the whole asthenosphere is 3.7×10^{11} erg, even more than the heating effect of the ^{40}K. This heating could readily be radiated away by the surface layers, however, so long as Th, U, stayed close to the surface. But if the heating effect of ^{40}K causes convection to set in, much of the Th, U, would indeed be carried downward, to depths from which heat could not escape except by convection. The effect would be considerably to reinforce the heating leading to convection.

It is important to realize that the heating effect of K, Th, U, would be quite insufficient to produce general melting at the Earth's surface, for then the radiation of the Earth into space would be vastly higher than the average heat generation by these elements. If the Earth were molten at its surface, with a temperature, say, of 1800 K, the radiation rate would be nearly 10^{35} erg year^{-1}, enormously greater than the average heat production of about 4×10^{28} erg year^{-1}. The surface must therefore be much cooler on the average than 1800 K. Indeed the 4×10^{28} erg year^{-1} from internal radioactivity is a thousand times smaller than the energy of the sunlight that falls on the Earth, so that the Earth can radiate the heat produced by radioactivity if on the average the surface is only a fraction of a degree hotter than it would otherwise be.

Since the surface material of the Earth must therefore remain overwhelmingly solid, one can wonder how the heat produced in the asthenosphere can actually reach the surface. A conceivable solution to this problem might be to have a thin solid crust with hot molten rock in convective motion extending below the crust to the base of the asthenosphere. Below the crust the temperature would then be about 1800 K, so that there would be a big temperature fall through the solid material of the crust. Such a temperature fall would cause a flow of heat by conduction, and if the thickness of the solid crust were properly adjusted this conductive flow would equal the rate of production of heat at lower depths by the K, Th, and U.

Such a situation would not be stable however. Lighter fractions separating from the molten rock, and accumulating in particular places, would be subject to pressure forces from surrounding material that would tend to force them upward through the solid crust, leading to the emergence of molten material at the surface.

And there would be a more fundamental instability, namely that the thickness of the solid crust could not be self-adjusting to the value needed for a steady equilibrium state. Thus if the crust were to get a bit too thin the convective flow of heat from below would be increased, which is contrary to the requirement for stability. This follows because the crust, being too thin, conducts heat more rapidly, tending to cool the material at the top of the convective columns. If this led to the solidification of the cooled material there would be a reduction of convection, because of the resulting large increase of viscosity, but solidification at the top would lead to just the same contradiction that we discussed in Chapter 12, a contradiction caused by the circumstance that the temperature gradient required to drive convection is less steep than the gradient of the melting curve. There would thus be an increase in the convective outflow of heat, preventing solidification, and tending to make the crust still thinner. And a too-thick crust would cause a reduction of the convective outflow of heat, leading to a still thicker crust. Hence the situation is not self-balancing with the right kind of feed-back.

A further possibility, which is closer to the resolution of the problem which the Earth actually adopts, is to avoid the separation of material into a solid static crust on the one hand and easily-moving fluid material on the other hand, by incorporating the crust itself into the convective motion. Since, however, the crust must be solid (in the conventional sense) we are then faced by a requirement for convection to occur in non-molten material. And in fact the asthenosphere is not molten, because if it were the S-waves discussed in Chapter 12 could not propagate there. The situation then is that the asthenosphere would

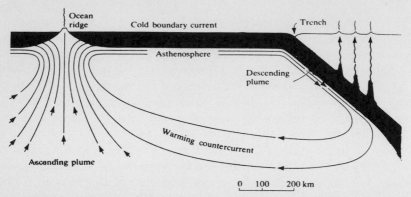

Figure 17.1. This form of circulation carries heat from the depths of the asthenosphere to the surface.

be fluid if it accumulated all the radioactive heat that is produced within it, but the asthenosphere is not fluid because it manages to rid itself of the radioactive heat by the kind of solid-convection illustrated in Figure 17.1. In this situation there is a non-isotropic situation with hot rock emerging along narrow ridges of comparatively small area (small enough areas so that the heat radiated by the hot rock is not too large) and with the rock dipping down again into the asthenosphere after it has cooled.

For many years the form of circulation shown in Figure 17.1 was thought to be impossible on the ground that the buoyancy forces developed by convection would be too small to cause any large scale motion of solid material. Recently discovered facts show, however, that motions like Figure 17.1 actually do take place. Because of the firmness of the new facts there has been a tendency to ignore the older arguments—which plainly were wrong at some point. But the old arguments certainly contain points of difficulty for the modern situation, or so it seems to me, and I will therefore return to discuss them in Chapter 19.

Chapter 18

The Surface of the Moon

If the Moon added as much K, Th, U, per unit area of its surface as the Earth did, then much the same situation described in the preceding chapter would have happened also for the Moon. But the amount per unit area of Th, U, added to the Moon was about $\frac{1}{4}$ of the amount per unit area added to the Earth. And the amount of K was probably significantly less also. If K were acquired from small particles spiralling towards the Sun through the Poynting–Robertson effect, then the K for the Moon would also be about $\frac{1}{4}$ of the amount per unit area for the Earth. And if, as is perhaps more likely, much of the Earth's potassium were acquired through the impacts of substantial bodies, the bodies (coming from at least the region of Mars if not from still greater distances) would have had encounter speeds that might well have been too high for similar impacting material to have been retained by the Moon. So the acquisition of radioactive materials was almost surely appreciably less per unit surface area for the Moon than it was for the Earth. Not only this, but the materials would have occupied a thinner outer layer than for the Earth. The possibility then exists that the heat generated by the radioactive materials could reach the surface of the Moon by conduction. There would then be no large rise of the sub-surface temperatures, and the lower rocks of the Moon would remain unmelted.

Observation shows that the process of Figure 17.1 has not occurred on the Moon. Nevertheless, almost all lunar geologists are of the opinion that molten rock emerged long ago at the Moon's surface from melting processes that took place in the sub-surface layers. The general concept is like that described in the preceding chapter for the Earth, the Moon being taken to have once processed a thin solid crust with molten material below. The molten material did not burst out on its own accord, however. The appearance of molten rock at the surface was precipitated by the puncturing effect of the impacts of substantial bodies from outside. In this way it is thought to connect the cratering of the Moon's surface, caused by impacts, with the flat floors which the larger impact areas on the front face of the Moon are observed to have, in particular with the flat floors of the maria. The floors are generally believed to be caused by the gushing out of molten rock from regions below the solid crust.

This picture would essentially have been proven if visits to the lunar maria had revealed the presence of extensive lava beds. But no astronaut has ever stood upon a lava bed. It is commonly stated that lava flows have been found on the Moon, but such statements are incorrect. Lava flows have been found only in the imagination of those who have examined distant photographs taken from the Apollo landing sites. What the astronauts actually found was a great deal of fine powder. Embedded or lying on the powdery surface was a multitude of rock fragments. Subsequent chemical analyses of the rock fragments showed them to be of highly variable chemical composition, making it quite unlikely that they were fragments from a single coherent flow of molten rock. Rather do they seem to be ejecta from impact events that occurred over wide areas of the Moon's surface.

The rocks have clearly been melted, fine-grained basalts having been recovered from the maria. But the profusion of small particles added to the Moon (and of which the Moon might even have been constructed) provides a medium of low sound speed for impacting objects, and this creates a highly favourable situation for impact-melting, quite apart from the possibility that the rocks are fragments of impacting objects which experienced melting before they hit the Moon. The heavily scarred lunar surface shows that there were very many impacts, often overlapping each other. Thus some regions would be melted and remelted. Impact areas would produce localized sheets of rock, which would then be available for break-up and scattering by subsequent impacts. And in encounters* of low speed (of under $2\,km\,s^{-1}$), the impacting objects could themselves have supplied rocky fragments to the Moon's surface. The isolated rock fragments do not demand lava flows from the interior of the Moon. Nor do rocks of a particular type, like basalt. Such rocks demand the fusing of the right kind of ingredients—in the case of basalt, ingredients with a high concentration of calcium oxide.

The importance of these considerations is that if we can establish that molten rock has not emerged from the interior of the Moon then there is a fair presumption that the interior of the Moon has not been molten. This would imply that the radioactive materials on the Moon were a late addition at the surface—they were never buried deeply. And similarly for the Earth we would infer that radioactive materials have never been buried more deeply than the asthenosphere, a deduction which would confine heating and convection to the asthenosphere, making it doubtful that the Earth is convective at deeper levels of the mantle.

Because of the great importance of these questions for terrestrial geology, I will now review a number of features of the Moon's surface, unconnected with the detailed study of rock compositions and structures, which seem to me to point overwhelmingly to a picture quite remote from that which is favoured by most lunar geologists. Or put it otherwise, if lunar geologists are to maintain their position in a satisfactory way, there is a whole array of other remarkable facts about the Moon which should surely be given convincing explanations, rather than being dismissed with clearly inadequate explanations, or with none at all. The point of view to be discussed from here on is one that has been emphasized and developed over the past two decades by Professor T. Gold.

A good place to start a discussion of these 'forbidden' issues is with the five huge impact craters found on the opposite side of the Moon, of which one is shown in Figure 18.1. The Moon always keeps the same side more of less toward the Earth, and it is the large impact craters on 'our' side that have the flat bottoms that we refer to as maria, as with Mare Imbrium in Figure 18.2. If the floor of Mare Imbrium is a lava flow, why is there no such flow on the floor of the big crater in Figure 18.1? Why is the phenomenon of the flat bottom shown not only by Mare Imbrium but by all large craters on 'our' side, but not by those

---

* The encounter speed is the speed of approach, not the impact speed. The latter is given by

$$(\text{Impact speed})^2 = (\text{Encounter speed})^2 + V_{esc}^2$$

where V_{esc} is the escape speed from the gravitational field of the Moon.

Figure 18.1. A large impact crater, comparable in scale to the maria on the near side of the Moon.

on the other side? How is it that big impacting objects hitting the Moon on our side could manage to punch through to underlying molten rock, whereas impacts on the other side did not do so? Because the protecting solid crust was thicker on the other side, some geologists have said. But the same phenomenon applies to craters quite a bit smaller than the biggest ones, showing that the difference between 'our' side and the other side would need to be large, the crust being much thinner on our side, to a point where the explanation becomes of very doubtful validity.

The same phenomenon is shown by the altimeter record of Figure 18.3. This record makes it clear that the difference of elevation between high ground and low ground is about twice as great on the far side of the Moon than it is on the near side. This difference is only explicable, I would suppose, if the low basins on the near side have been submerged by some kind of 'filler' material, by a sea of material out of which only the high ground rises. Although this explanation might at first sight be thought to admit lava as a possible filler material, how then do we explain another important feature of the altimeter records, namely, that there are far more sharp peaks on

Figure 18.2. The near side of the Moon, with Mare Imbrium at bottom right.

the far side than on the near side? To explain this remarkable difference we have to suppose that the filler material has been able to smooth the high ground on the near side as well as fill up hollows in the low ground. How do lava flows smooth high ground? A fine powder, such as the astronauts have always found on every lunar landing site, a fine powder able to compact itself into a friable rock under pressure, would seem to me a far better candidate for the filler material than solid lava. Fine powder could fill out irregularities in the high ground, whereas lava could not achieve this at all.

But why should there be a drastic difference in the behaviour of fine powder on the two sides of the Moon? Let us see if we can outline an answer to this critical question.

The Moon moves for about 4 days in each synodic month of 29.53 days (new moon to new moon) into a region where the Earth's magnetic field is affected by the wind of particles which the Sun emits all the time. During these 4 days the Earth is lying more or less between the Moon and the Sun, so that the near side, 'our' side of the Moon, is turned toward the Sun at this time. The interaction of the

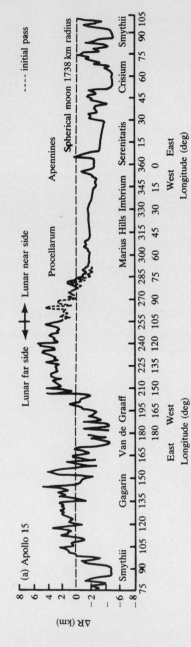

Figure 18.3. An altimeter record which shows that the difference of elevation between high and low ground is about twice as great on the far side of the Moon than it is on the near side.

Earth's magnetic field and the solar wind is known to produce fast electrons, some of which will inevitably strike the Moon's surface. Professor T. Gold has suggested that such electrons hit the Moon's surface preferentially from the direction of the Sun, in which case they would impinge only on 'our' side of the Moon, not on the 'other' side.

If this is so, then it becomes possible to understand how fine powder can move by large distances over the near side of the Moon, whereas on the far side a similar powder will remain more or less fixed at the place where it originally fell upon the surface. Vacuum experiments in the laboratory, using a powder with grains of the size that lunar powder is known to have, show a curious phenomena when electrons with bombarding energies in the range from 300 to 600 V impinge upon it. Grains at the surface then hop around vigorously like a swarm of fleas. According to Professor Gold, this observed phenomenon occurs when the incident electrons strike out on the average one secondary electron for each incident electron. The ejection of secondary electrons depends on the energy of the incident electrons. When the latter is small there are few secondary electrons, and the powder is stable, with the grains all tending to charge themselves negatively through the addition of the incident electrons. And for very high incident energies the powder is again stable, with the grains all tending to become positively charged through the escape of excess secondary electrons. With the charge on each grain of the same sign, no particular grain can acquire much of an individual charge, however, otherwise the whole powder would collectively acquire too great a total charge. (The whole powder acts collectively to draw ambient positive charge to itself when the incident electrons have low energy, and to draw negative charge in the high energy case.) And because individual grains cannot have much in the way of separate charges, there are no strong electrical forces between them.

The situation is otherwise when the number of secondary electrons average to equality with the number of primary electrons. For some particles, depending on shape and on composition, there will be slightly fewer secondaries ejected than there are incident electrons, and these particles will become negatively charged, while for other particles the opposite situation will occur. Each particle is balanced on a hair-trigger, and some go positive while their neighbours go negative. Positive and negative charges are therefore well-mixed and there is no problem about the whole powder collectively acquiring a large charge of either sign. In these circumstances, neighbouring particles can acquire large individual charges, and the resulting electrical forces between them are also large. According to Professor Gold it is these forces which cause the particles in his experiments to hop around with such remarkable vigour.

Professor Gold considers that grains can in this way develop individual electrical potentials that are comparable with the energy of the primary electrons, say, 500 V. Suppose that the resulting electrical force between a particle and its neighbours caused it to come free of the entanglements of other grains and to leap out of the surface. How fast might the electrical forces cause it to move? For particles of size 1 micron $(10^{-4}$ cm), the resulting speed would be about 0.1 m s^{-1}, sufficient in the low gravitational field of the Moon to permit a particle to hop by a horizontal distance of a centimetre, sufficient when repeated many times, over many millions of years, to permit an effective horizontal migration of fine powder.

In any part of the Moon's surface with a slope, the nett effect of gravity on a particle which makes many hops will be to cause it to move systematically downhill. Thus the trend must always be for particles to fill in hollows, and this will be true no matter where the hollows occur, whether for hollows on the high ground or for hollows on the lowest ground. The particles that compose friable walls of craters will tend to flow to the bottom of the craters, building flat bases of the kind that are observed in all the larger craters that happen to be situated on the near side of the Moon. Should the wall of a crater on low ground chance to be breached at any place on its circumference there will be a flow of particles into the crater from outside. Such in my view is the process that has produced the flat bottoms of the maria. This view in spite of the evidence (of which there is more to come) is not popular, and in joining Professor Gold in holding it I am conscious of placing myself very nearly in a minority of two.

A lump of rock ejecta falling on a mare surface would initially dig a hollow for itself. The rock shown in Figure 18.4 is not seen lying in a hollow, however. Powdery material has flowed around it, just as powdery material was found to do in the laboratory experiments (Figure 18.5). The laboratory experiments showed that initially separate powdery materials of different chemical compositions remain immiscible. One material can be made to flow across another, and in such a situation the boundary between the materials is always sharp. Sharp boundaries are to be seen very frequently in detailed pictures of

Figure 18.4. The rock shown here is not lying in a hollow. The scale is about half a metre (from Apollo 12).

the Moon's surface, as in Figure 18.6, while Figure 18.7 shows a crater in the process of being overwhelmed by a flow of powder.

The initial situation resulting from a large impact on the lunar surface would be like the big crater to be seen on the far side of the Moon, like Figure 18.1. The impact would have had two effects, (i) to cause spill-out and scattering of surface material from the area of the crater, and (ii) to compress and perhaps to fuse lower-lying material. The scattering and spill-out must cause the area of big craters like Figure 18.1 to have negative gravity anomalies—gravity must be slightly weaker over those areas because of the loss of the scattered material. Now because of (ii) there will be an excess lowering of the floor of the crater—quite apart from the lost material—since the remaining material becomes compressed and therefore occupies less

131

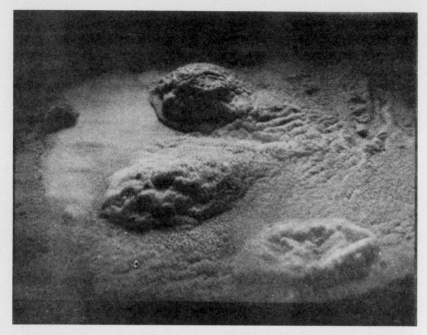

Figure 18.5. In a laboratory experiment with fine powder the powder flowed around several obstacles, forming a sharp boundary at the base of the obstacles, as in Figure 18.4 (courtesy, Professor T. Gold).

volume than it did before. So if powder with the normal average density of the Moon's surface layers flows into the crater and manages to refill the excavated volume, the gravity anomaly will be changed from being negative to becoming positive. The filled craters on the front side of the Moon, like Mare Imbrium in Figure 18.2, show positive gravity anomalies (mascons) while the big craters on the far side show negative anomalies. It defies commonsense to argue that a lava flow would produce a positive gravity anomaly, since a fluid flow would surely adjust itself to zero anomaly, i.e. to isostasy.

There are no internal movements of rock inside the Moon that are at all comparable with those which happen frequently in the outer layers of the Earth. Internally-generated moonquakes are more than 1,000 times weaker than earthquakes. The largest moonquakes occur when the Moon experiences impact from meteorites, or even from a rejected lunar-landing vehicle thrown back upon the surface. Whereas the seismic signals from earthquakes are brief, the signals recorded by

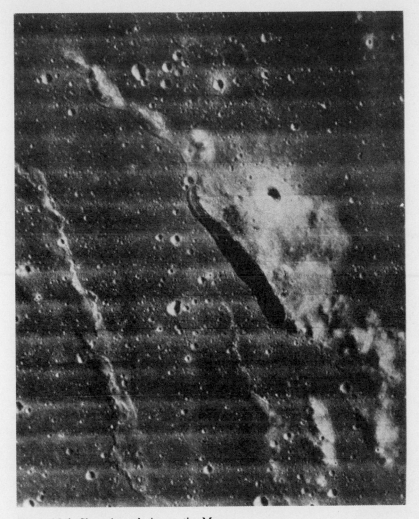

Figure 18.6. Sharp boundaries on the Moon.

seismographs landed on the Moon show that the vibrations of moon-quakes last without any large decrease of amplitude for upwards of an hour. This is decisive evidence that the surface layers of the Moon are quite different from the surface layers of the Earth.

The long persistence of the vibrations caused by moonquakes can be explained by a deep layer of powder gradually compacting itself with

Figure 18.7. An advancing sea of powder flowing over a crater.

depth. It may also be possible to explain the seismic observations if the Moon's surface were made up of small broken rocks and rubble, with only a thin surface skin of fine powder. But this second possibility (so far as the seismic evidence is concerned) is sharply contradicted by evidence of a quite different kind. If there were only a thin skin of powder with broken rock underneath, then the Moon would scatter longwave radio signals from the Earth over a wide angle. And different

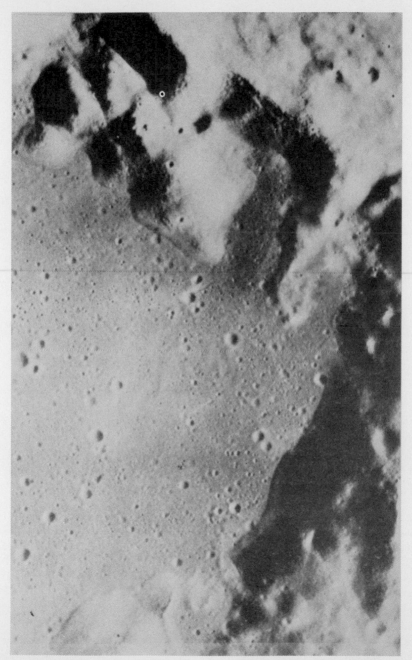

Figure 18.8. The filler material does not push down bits of crater walls that emerge from it.

parts of the Moon's surface would return such signals to the Earth in a rather uniform way. Where one looks at the Moon with the eyes, the surface appears more or less uniformly bright—there is comparatively little 'limb-darkening'. This is because for ordinary light the Moon is everywhere rough—there is always some surface at any place on the Moon's surface that is appropriately oriented to reflect sunlight towards the Earth. And if broken lumps of rock were the essential characteristic of the lunar surface, the same would be true for radio signals, which emphatically it is not. Radio signals are reflected 1,000 times more strongly from the centre of the lunar disk than they are from the limb. A comparable situation optically would demand a smoothly polished ball, so smooth that in shining light on it one would see only a small central bright spot. A deep layer of powder is consistent with this further observation, since for radio waves such a surface would indeed behave as a polished ball does optically. The observation is quite inconsistent, however, with the broken-rocks picture.

There are many examples of only a portion of the wall of a crater sticking up above a floor of the filler material, as in Figure 18.8. The filler material has almost covered over such craters and yet it has contrived not to push down the fragile bit of the projecting crater wall. This is understandable if the filler is an even more fragile powder. It is hardly understandable for a flow of lava.

This is not all. There are other quite different lines of evidence, evidence from the uniformity of the track densities of cosmic rays to be found in fine particles recovered in cores taken from the lunar 'soil'—a word coined by lunar geologists to avoid saying 'powder' or 'dust'. Contrasting with this uniformity of cosmic-ray track densities, there are remarkable variations of chemical composition between one layer of such core samples and another, composition variations which arose not from a magmatic differentiation on the Moon but from the time sequence in which the particles were added to the Moon in the first place. The evidence all points systematically towards the conclusion that the Moon's surface as we see it today is an accumulation surface—the Moon was formed much the way we see it now.

Chapter 19

Continental Drift and Plate Tectonics

The common element magnesium is deficient in the rocks of the Earth's surface. In the original material of the solar nebula magnesium was as abundant as silicon, and yet at the Earth's surface magnesium is about six times less abundant than silicon. This remarkable situation seems to be due to enstatite rather than forsterite being the principal stony material going to form the mantle of the Earth—or at any rate the outer part of the mantle. Then as a consequence of melting one can have the separation

$$2MgSiO_3 \rightarrow Mg_2SiO_4 + SiO_2.$$

The curious aspect of this geochemical separation is that while enstatite and forsterite have nearly the same density, $3.19 \, g \, cm^{-3}$ for enstatite and $3.21 \, g \, cm^{-3}$ for forsterite, silica in the form of quartz has a density of only $2.65 \, g \, cm^{-3}$, and other forms of silica have still lower densities. Silica is the principal component of granite (about 60 per cent), and granite is the principal component of the rocks which form the continents of the Earth. The formation of the continents seems therefore to have been based on the above separation of silica from enstatite, the mass of continental rock being determined by the quantity of enstatite to have undergone this transformation.

The lighter rock of the continents has been aggregated into a number of pieces each of which has experienced the phenomenon of isostatic compensation. According to Chapter 12 and to the discussion of Chapter 17, the Earth was mainly accumulated by impacts of considerable magnitude, each of which produced an added layer that was fluid for a short while following the impact. The inability of a fluid to withstand stress causes it always to adjust itself so that the surface is a gravitational equipotential—there is no component of the gravitational force along the surface, only at right-angles to the surface. The accumulation of the Earth therefore proceeded in an isostatic manner—when the material solidified, the pressure (or more accurately the stress-tensor) was essentially without variation along the equipotential surfaces.

The formation of the continents with a vertical thickness of some 35 km could have changed this situation, but it did not do so except in a minor degree. As the continents accumulated over particular parts of the Earth's surface there must have been an outflow of underlying material which continued until the downward force of gravity was very nearly the same everywhere over the surface, over continents and oceans alike. The present-day surface differs little from a gravitational equipotential, on the average by only a height difference of about 30 m. Since the weight of 30 m of rock is about 10^4 g, and gravity is 981 c.g.s. units of acceleration, the pressure at the base of 30 m of rock is about 10^7 c.g.s. units (about 10 bar). Thus the implication is that there are no horizontal pressure variations in the asthenosphere of more than about 10 bar. Although the pressure exerted by a whole continent is as much as 10,000 bar, isostatic compensation due to the outflow of material below the continents has reduced the horizontal pressure variations to about 1/1,000 of what they otherwise would have been.

The relevance of these considerations is that they give a limit of about 10 bar to the horizontal pressure variations which can drive the convective motion of Figure 17.1, reproduced here as Figure 19.1. This already sets a serious problem for an application that has been widely suggested and accepted for the effect of this circulation. Thus quite apart from the purpose of Chapter 17, which was to transfer heat from the asthenosphere to the surface, it is usually supposed that the uplift of mountain plateaus and the buckling of mountain chains are to be attributed to the same circulatory process. But the latter phenomena can be caused only by horizontal pressure variations that are larger than the yield strength of rock, and the yield strength exceeds 1,000 bar. The discrepancy here is so large as to suggest that the circulation of Figure 19.1 is not responsible for the uplift of

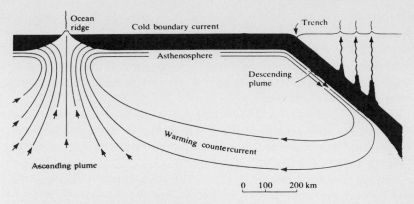

Figure 19.1. This form of circulation carries heat from the depths of the asthenosphere to the surface.

mountain ranges or for pushing up high ground like the plateau of Tibet. And if some other process is responsible for generating the much larger stresses needed by these phenomena then one can legitimately wonder whether the circulation of Figure 19.1, if it occurs, is really a primary process.

If we write P for the magnitude of the horizontal variation of the pressure in the asthenosphere, then the speed u of a circulatory motion of the type of Figure 19.1 is related to P and to the dimension l of the motion by the relation

$$u \approx \frac{Pl}{\eta}, \tag{1}$$

where η is the effective dynamical viscosity of the material. The dimensionalities of u, P, η make this relation inevitable. Thus u has dimensionality (length) × (time)$^{-1}$, P has dimensionality (mass) × (length)$^{-1}$ (time)$^{-2}$, while η has dimensionality (mass) × (length)$^{-1}$ (time)$^{-1}$, and (1) is the only relation between these quantities that has the same dimensionality on both sides of the equation. So unless some other physical quantity enters the problem (which it does not) this is the only way in which $u, l, P,$ and η can be related. A numerical factor, like π, or like the dimensionless ratio of different dimensions of the system, can appear but nothing more than that. For instance, the ratio of the horizontal dimension of Figure 19.1 to the vertical dimension is of order 10, and this factor could enter an equality relation between u and Pl/η. Significant as such an order of magnitude factor might seem, it is not as important as the value to be attributed to η, which can change by many orders of magnitude, even for temperature variations of only 100 K. As material goes fluid η changes by more than 10 orders of magnitude (more than 10^{10}). So the value to be used for η in (1) is of greater importance than any numerical factor we may have omitted in this expression.

Figure 19.2 shows the kind of variation of η with depth usually employed by geologists and geophysicists. As an average representation of η we may therefore choose $\eta = 3 \times 10^{22}$ g cm^{-1} s^{-1} (poise). With $l = 3,000$ km s^{-1} for the horizontal dimensions of Figure 19.1, and for $P = 10$ bar (i.e. 10^7 g cm^{-1} s^{-2}), (1) gives $u \doteq 10^{-7}$ cm s^{-1}, or about 3 cm year^{-1}, which is close to the circulation velocity considered desirable for the system of Figure 19.1.

Such velocities are actually obtained from the observed phenomenon

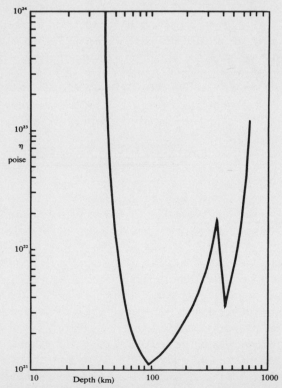

Figure 19.2. Schematic relation between viscosity and depth.

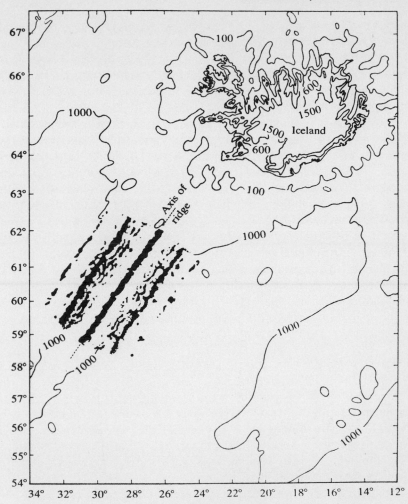

Figure 19.3. Molten rock emerging from a ridge becomes magnetized as it cools. Spreading on both sides of the ridge forms patterns reflecting variations in the Earth's magnetic field. The eastward and westward patterns are remarkably similar.

of sea-floor spreading, through the correlation of magnetic patterns of the sort shown in Figure 19.3. Hot rock emerging both eastward and westward from the mid-Atlantic ridge becomes magnetized as it cools. The form taken by this magnetization changes with time, because the Earth's magnetic field which induces the magnetization itself changes with time. In the representation of these time variations, the rocks

141

flowing to the east mimic in their magnetization that of rocks flowing to the west, an effect shown dramatically by the correspondences between the eastward and westward patterns in the figure. With estimates for the time-scale of the changes in the Earth's field independently available, one can easily estimate the rate at which the patterns of Figure 19.3 must have spread apart from each other, a rate for spreading from the mid-Atlantic ridge of about 1.25 cm year^{-1}.

But of course this unquestioned observation does not establish the emergence of hot rock from the mid-Atlantic ridge as the primary aspect of the circulation of Figure 19.1, as some authors apparently suppose. Nor does it prove that heating due to radioactivity is the cause of the circulation, although another simple calculation is often considered to give decisive support for this interpretation. An estimate for P can be obtained from a consideration of buoyancy forces which generate horizontal density variations $\alpha\rho\,\Delta T$ where α is the volume coefficient of expansion, ρ is the average mass density, and ΔT is an assumed horizontal temperature difference between rising and falling material. Such horizontal density variations maintained over a height interval h would produce a horizontal pressure variation P given by

$$P \approx \alpha\rho\,\Delta T \times gh, \tag{2}$$

where $g = 981$ cm s^{-2} is the acceleration due to gravity. Putting $\rho = 3$ g cm^{-3}, $h = 100$ km, and taking α to be 4×10^{-5} per degree kelvin, (2) gives $P \approx 10^{6}\,\Delta T$ g cm^{-1} s^{-2}, with ΔT in degrees kelvin. Thus a systematic horizontal temperature variation of about 10 K would generate a pressure difference $P \approx 10^{7}$ g cm^{-2} s^{-2}, sufficient according to (1) to generate a speed u of the order of the observed sea-floor spreading values. Such a temperature variation is usually considered to be entirely reasonable.*

So there is a general consistency about the argument concerning the circulation of Figure 19.1, and no doubt this consistency explains why at least some of the features of this circulation actually occur, as one can evidently see from Figure 19.3. Difficulties emerge, however, when we consider the relation of (1) to the viscosity curve of Figure 19.2. The viscosity varies from very high values above 10^{24} g cm^{-1} s^{-1} close to the surface to about 10^{21} g cm^{-1} s^{-1} at a depth of 100 km s^{-1}. Maintaining a coherent value of the circulation velocity u in (1)

---

* The vertical temperature variation over a height range of 100 km would be about 40 K for a purely adiabatic situation, so that a superadiabatic temperature gradient about 25 per cent greater than the adiabatic gradient would produce the required ΔT.

therefore requires a much higher pressure near the surface than is necessary for the lower region of smaller viscosity. Indeed the value of P needed to maintain $u \simeq 3$ cm year^{-1} (such as is required at the surface by sea-floor spreading) exceeds 100 bar for the top 50 km of material. The strange aspect of this result (which has been arrived at in a number of extensively argued papers by geophysicists) is that the pressure forces needed to drive the circulation are very much larger close to the surface than they are in the depths of the asthenosphere. The implication of this curious situation is that the circulation, if it occurs in the complete sense of Figure 19.1, cannot be driven simply by the generation of heat at depths in the manner of a saucepan heated from below. The driving agency is required either to lie near the surface, or the process differs from anything we have considered so far.

It has been suggested that large values of ΔT should be used close to the surface. Rock at the surface cools to comparatively low temperatures. Should such rock sink without picking up much heat from surrounding hotter material, a large value of ΔT will be generated. With $h = 50$ km, $\Delta T = 500$ K, (2) gives $P \approx 3 \times 10^8$ g cm^{-1} s^{-2}. In this picture, sea-floor spreading occurs, not from an excess pressure forcing molten rock up through the oceanic ridges, but through suction produced by falling cooled rock, an inversion of the usual saucepan-heating situation.

My own view of the matter is somewhat similar in effect to this last picture, but it differs in principle. The density variations between different rocks, as for instance the density difference between silica and forsterite, are much bigger than the density variations occasioned by temperature variations. Density variations produced by local differences of chemical composition or of separation, as with $2MgSiO_3 \rightarrow Mg_2SiO_4 + SiO_2$, can lead to buoyancy forces much bigger than (2). For a lump of heavy material of density ρ_1 falling through lighter material of density ρ_2, one has in place of (2),

$$P \approx (\rho_1 - \rho_2)gh. \tag{3}$$

Putting $\rho_1 = 3.2$ g cm^{-3}, $\rho_2 = 2.6$ g cm^{-3} and taking h as small as 10 km, gives $P \approx 5 \times 10^8$ g cm^{-1} s^{-2}. Indeed the effect of even much smaller density variations than this would easily dominate the thermal variations. Comparing (2) and (3), the relative importance of the two sources of P is determined by $(\rho_1 - \rho_2)/(\rho_1 + \rho_2) \div \alpha \Delta T$. Even for ΔT as high as 1,000 K, (3) gives as large a value of P as (2) using the same h with only a 10 per cent difference between ρ_1 and ρ_2. Unless we are prepared to argue that the material of the asthenosphere and of the Earth's crust (lithosphere) are very homogeneous in their composition density variations will evidently exert a controlling influence over thermal convection.

The heat produced by radioactivity still plays a critical role, however, in promoting the separation of the different density fractions and by increasing the temperature sufficiently to reduce the viscosity η to values at which the heavier density fractions can fall down through the asthenosphere.

As a heavy lump falls, the resulting pressures act to cause a horizontal inflow of material above the lump and an outflow of material below the lump. Material falling, not just in one place, but below a line drawn over the surface of the Earth, would thus generate suction toward the line at upper levels and outward pressure at lower levels, which is the kind of driving agency required by the circulation of Figure 19.1. Such lines, again in my view, are generated by the positions of the continental masses. Uranium and thorium have become especially concentrated in the granitic rocks of the continents, which therefore act like hot pads sitting on the top of the asthenosphere. The continents are thought, moreover, to be growing in their masses through a continuing addition of a low density fraction high in its silica content. The continents may well be a means of sorting the silica fraction (density about $2.6 \, \text{g cm}^{-3}$) from the so-called 'ultrabasic' fraction rich in MgO, with the latter falling to the bottom of the asthenosphere over areas that are covered by the continents. If a continental mass happens to be more or less linear in shape, as the Americas are, a line of fall-out is defined which runs the length of the continent.

Once a falling lump reaches a depth of about 100 km the value of η falls to about 10^{21} poise, and the speed of fall increases considerably in accordance with (1). The lump falls comparatively quickly to the base of the asthenosphere. Thus the pressures generated by dense lumps of material act longest when the lumps are in the region of high viscosity not too much below the surface, the length of time for which they are able to generate the pressure variation P being determined by the degree of heating which they experience, the heating which is required to produce some reduction of the exceedingly high surface values of η before falling can take place.

The values of P given by (3) apply in the immediate vicinity of a heavy lump of material. If the lump is large, a comparable pressure variation will extend to considerable horizontal distances, but if a lump is small the pressure variation will decline considerably at large horizontal distances from the lump. Thus a lump may generate $P \approx 1,000$ bar in its own vicinity but only, say $P \approx 10$ bar at large distances from itself. Another way of expressing the dominance of inherent density variations of rock types over density variations of purely thermal origin is to say that the amount of the falling high density fraction required to produce, say, $P \simeq 10$ bar over large distances, is much less than the amount required for the purely thermal case. Indeed, 10 to 100 times less, depending on what one assumes for ΔT in the thermal case. This consideration of amount is important because, unlike the thermal case, the same chunk of rock cannot be used many times—once a heavy lump has fallen to the bottom of the asthenosphere it does not rise again.

We noted above that the surface rocks are very deficient in MgO. If one were to add MgO to the continents until their magnesium content equalled the silicon content (as in the original condensates from the planetary gases) the amount of added material would increase the continental mass about 40 per cent. And if one were to do the same for an additional layer 10 km thick covering the whole Earth the further amount needed would be about the same—another 40 per cent of the continental mass—giving a total addition not much different from the total mass of all the continents. And if one somewhat arbitrarily divides the lifetime of the Earth, 4.6×10^9 years, into 'cycles' of 2.3×10^8 years, there have been 20 such 'cycles', and the amount of heavy material available on the average for fall-out in each cycle is about 5 per cent of the total mass of the continents. To compare with the thermal case, this 5 per cent must be multiplied by a large factor, a factor of 40 for $\Delta T = 100$ K in the thermal case, and the value of P which can be generated over large horizontal distances is given by putting $T = 100$ K in (2). With h taken as 100 km, we then get $P \simeq 10^8 \text{ g cm}^{-1} \text{s}^{-2}$. Taken as an average, such a value of P is adequate to drive the system of Figure 19.1, the 'cycle' time chosen above being of the order of the circulation time of the material in the figure.

But the pressure variations close to particular lumps are larger than the average value estimated in the preceding paragraph. And the heavy lumps will not always fall at a steady rate. Nor will they fall uniformly everywhere over the Earth. We saw above that fall is likely to be concentrated under the hot pads of the continents. Thus immediately under the continents the horizontal pressure variations will be larger, and will be directed inwards, tending to compress the continents into smaller and thicker blocks.

Some such compressive force appears to be necessary to combat erosion effects acting on the raised land of the continents. Erosion happens quickly in a time-scale of a few million years, and unless effectively opposed by a strong compressive effect would soon wear down the high ground, spreading out the continental margins through sedimentation. The compressive force acts through (3), which with $h = 100\,\text{km}$, $\rho_1 = 3.28\,\text{g cm}^{-3}$, $\rho_2 = 2.6\,\text{g cm}^{-3}$, gives $P \approx 5 \times 10^9\,\text{g cm}^{-1}\,\text{s}^{-2}$. Such high values of P, applying locally close to the vicinity of sinking material, exceed the yield strength of rock and would be capable of producing the mountainous regions of the continents.

We remarked above that heavy material falls quickly through the region of low viscosity in the asthenosphere, reaching a base where the value of η rises steeply again. I would expect the heavy material as it reaches this base to spread out horizontally into a layer distributed effectively uniformly over the Earth. Isostatic compensation over areas of exceptional fall-out then requires an inflow of material from surrounding regions, an inflow that will take place most readily at a depth where η is least, according to Figure 19.2 at a depth of about 100 km. The trend is thus toward a density sorting, not just of material close to the surface, but of the asthenosphere itself. At the base of the asthenosphere under the continental band of the Americas we have a smoothing outflow of heavy material compensated isostatically by the inflow of lower density material into the regions of the asthenosphere immediately below the continental material. This accumulation must cause the asthenosphere to lift, creating an upward bulge, tending also to lift the continental material. So as well as a generally compressive force on the continents there is also a tendency for lift because of the asthenospheric bulge which develops along the line of the continent, a bulge which may form itself into a ridge if the continental mass is sufficiently linear in shape.

Figure 19.4 shows the world-wide system of rifts at which the circulatory pattern of Figure 19.1 is thought to emerge and to descend. The relevant aspect of Figure 19.4 for the present discussion is the remarkable similarity of shape of the mid-Atlantic ridge to the eastern side of the Americas and to the western side of Europe and Africa. It seems clear that at one time (thought to be about 2×10^8 years ago) the mid-Atlantic ridge formed as a bulge which lifted itself below a larger continent combining Africa and South America and combining Europe

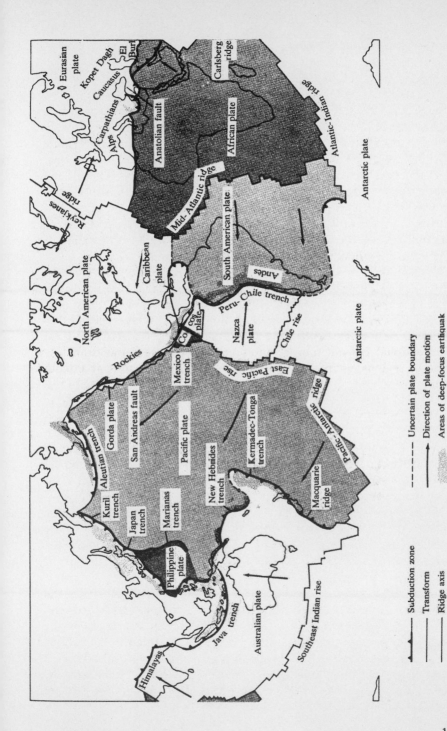

Figure 19.4 The distribution of plates in the Earth's lithosphere, with motions indicated by the arrows.

and North America, *and that the bulge lifted itself more or less along the central axis of that larger continent.* The lift must have placed the continental rock under tension, and rock is about 10 times weaker under tension than it is under compression. So for a lift effect that was quite a bit weaker than the compressional forces required to produce the buckling of mountain chains, the combined continent would have been cracked open, a division thus forming along its central axis. The process would be essentially catastrophic, because once a gap was forced open the asthenosphere below the gap, being then freed of the weight of the previously overlying continent, would bulge up still more drastically. A high ridge would thus appear as the two halves of previously-joined continent became forced apart from each other.

On this point of view it was the shape of the previously-existing larger continent that determined the shape of the mid-Atlantic ridge, not vice versa, as often seems to be supposed. The shape was defined by the heating pad effect which lowered η values below the combined continent, giving a more rapid fall there than elsewhere for the heavy density fractions of rock. This in turn led to an accumulation of a lower density fraction below the larger continent (as the heavier density material fell to the base of the asthenosphere and spread itself outward from there) and it was this accumulation which led to the formation of the ridge itself. The phenomenon develops therefore from regions of high η to regions of lower η, as it must surely do. This indeed seems to me the main lesson of continental drift and plate tectonics—cause and effect must go in the sense from higher η to lower η. It cannot go the opposite way, as frequently seems to be imagined.

Chapter 20

The Earth's Initial Supply of Carbonaceous Material

We have considered two ways in which carbonaceous material could have reached the Earth, a direct component from the solar nebula, and an indirect component acquired from the interstellar cloud in which the solar system was born by the apple-corer process discussed in Chapter 8. The detailed composition of the latter has so far been thought to be represented by the molecules set out in Table 8.1, while the former has been taken to be CO_2, resulting from

$$CO + H_2O \rightarrow CO_2 + H_2,$$

an exothermic reaction that became thermodynamically preferred as the temperature within the solar nebula fell below about 600 K. In this final chapter two possible modifications to these former considerations will be discussed.

While a thermodynamic calculation determines the most stable molecules (for a given temperature and for given atomic concentrations) there is no guarantee that a non-thermodynamic initial situation will relax to the thermodynamic state within the available time-scale. Reactions like the conversion of CO to CO_2 may be too slow for the thermodynamic state to be achieved. If this were so, then the Fischer–Tropsch reaction

$$CO + 3H_2 \rightarrow CH_4 + H_2O$$

would require discussion. However, the Fischer–Tropsch reaction is strongly pressure dependent. At the pressure within the planetary gases in this region of Uranus and Neptune, $\sim 10^{-6}$ atm, it seems unlikely that much CH_4 would be produced. Hence carbon in the solar nebula would be likely to remain as CO, if the production of CO_2 should be too slow. Moreover, the CO would remain gaseous. It would not condense into the multitude of icy bodies in the region of Uranus and Neptune (the melting point of solid CO is $-199°C$, compared to $-56°C$ for solid CO_2). Nor would the CO, because of its high molecular weight, be evaporated from the periphery of the solar system, like H and He were. The CO would be added eventually by gaseous accretion to Uranus and Neptune, where at the high pressures in the atmospheres of those planets, and with some H also present, the Fischer–Tropsch reaction would lead to the production of CH_4. But this would be on Uranus and Neptune, not on the Earth, or on the other terrestrial planets. Thus the effect of CO not being converted to CO_2 by the first of the above reactions would be to cut-off the supply of carbonaceous material from the solar nebula to the terrestrial planets. Only carbonaceous material from the interstellar cloud would then have reached these planets, first by being picked up by the apple-corer process of Chapter 8, second by condensing onto the swarm of icy bodies in the region of Uranus and Neptune, and third through a small fraction of those bodies being perturbed into highly elongated orbits, causing a still smaller fraction eventually to collide with the inner planets.

The molecules of Table 8.1 may be far from representing the full measure of complexity of the composition of interstellar material. Indeed, many of the molecules of Table 8.1 may be no more than degradation products, resulting from the break-up of far larger molecular structures. In support of this contention, Professor Wickramasinghe and I have shown,[*] I think with a fair presumption of being correct, that the organic material within the interstellar clouds may have a measure of complexity comparable to that which occurs in biological systems. Quite a number of the interstellar clouds contain within themselves compact sources of infrared radiation, which under spectroscopy show a wide range of detailed patterns. Astronomers are satisfied that for many of these sources the emitting material is 'dust'—that is to say, small solid particles of some kind. The dust can also be observed at ordinary visual wavelengths, at which wavelengths it blocks starlight, as in Figure 20.1 (see Plate IV in the colour section).

---

[*] F. Hoyle and N. C. Wickramasinghe, *Nature*, **268** (1977), 610.

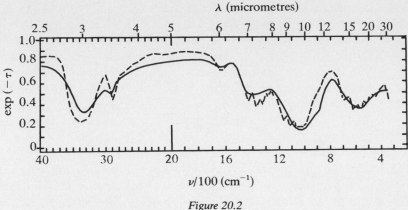

Figure 20.2

Professor Wickramasinghe and I tried to match the laboratory absorption spectrum of the commonest of all biological molecules, cellulose, to the observed astronomical sources. The laboratory spectrum for cellulose is shown by the broken curve of Figure 20.2. The solid curve in this figure contains two modifications from the measured cellulose curve. The minimum at 3μ is shallower, and there is an absorption excess at about 11μ. The shallower minimum represents our estimate of what unwetted cellulose would give—laboratory cellulose always contains free water, which astronomical cellulose would not. The modification at 11μ was made to include the effect of simple hydrocarbons like C_3H_6, the presence of which seems to be clearly indicated in some of the observed sources.

Figure 20.3 shows the emission at a temperature of 175 K expected for our modified spectrum, together with points representing observations for the astronomical object the Trapezium Nebula. The agreement is very good even to wavelengths as long as 30μ.

Sometimes the 'dust' acts in absorption, rather than in emission as in Figure 20.3. This happens when a cool dust cloud intervenes between ourselves and a distant source of infrared radiation. An example of dust acting in absorption is shown in Figure 20.4. Once again the curve has been calculated for cellulose (more accurately for the modified solid curve of Figure 20.2) and the points are the astronomical observations. This case for the source OH26.5 + 0.6 is particularly remarkable for its wide range of wavelength, 2μ to about 35 μ, and for the ability of the calculations to follow the observations over such an extended range.

151

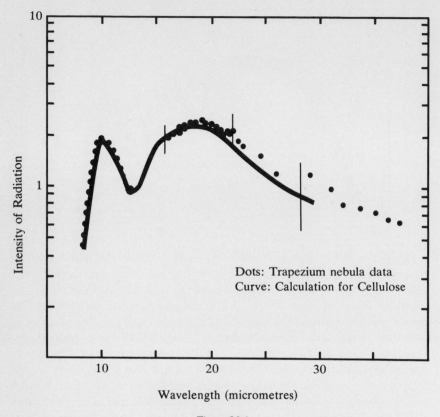

Dots: Trapezium nebula data
Curve: Calculation for Cellulose

Wavelength (micrometres)

Figure 20.3

There are still more complex cases, of which Figure 20.5 is an example. This gives observations for the infrared source BN, named after Drs. Becklin and Neugebauer who first observed it. The 'dust' in this object acts both in emission and absorption. Once again, the solid curve of Figure 20.2 leads to a calculated curve that is in good agreement with the observed points.

Two kinds of material are known to astronomers. The more common contains hydrogen in great excess, oxygen in greater concentration than carbon, and nitrogen in comparatively low concentration. It would be in this type of 'normal' material that cellulose might be

---

* F. Hoyle and N. C. Wickramasinghe, *Nature*, **270** (1977), 701.

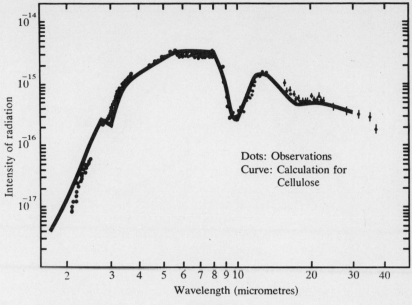

Figure 20.4

expected to form.* The less common kind also contains hydrogen in excess, but nitrogen in greater concentration than carbon, and oxygen in low concentration. Nitrogen atoms would replace oxygen in rings similar to those of cellulose (pyridine rather than pyran rings), while bonding between rings would pass through NH_2 groups rather than through oxygen atoms. The second kind of material is produced whenever the first kind is subjected to the carbon–nitrogen cycle of nuclear reactions within stars.

There is no firm evidence from the infrared observations for the existence of heterocyclic nitrogenated rings, but there is evidence in the visual and ultraviolet parts of the spectrum. Light from bright stars often shows several broad absorption bands caused by the interstellar material through which the light passes on its way to the Earth. There is a particularly strong feature centred at a wavelength of 4,430 Å ($1Å = 10^{-8}$ cm). Attempts to explain this feature in terms of absorption by common inorganic molecules have failed, but some years ago F. M.

* F. M. Johnson, 'Interstellar Molecules and Cosmochemistry', *Ann. N.Y. Acad. Sci.*, **194** (1971) 3.

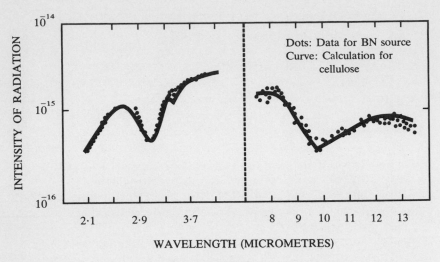

Figure 20.5

Johnson pointed out that the porphyrins (four C_4N rings joined through the same atom, Fe, Mg, . . .) have a marked absorption near this wavelength.* (Magnesium–porphyrin forms the basic part of the chlorophyll molecule.) Johnson's suggestion has not been taken very seriously by astronomers, but perhaps in view of the apparent relation of cellulose to observations in the infrared it will now receive the attention which I suspect it deserves.

There is an absorption of starlight in the ultraviolet at 2,200 Å, which coincides with absorption of the doubly nitrogenated ring C_4N_2, joined to a subsidiary carbon ring, as in quinazoline. While this coincidence, if it stood alone, would hardly persuade one to believe in the widespread occurrence of quinazoline, taken with the cellulose situation it assumes a fair measure of strength. Not only this, but a C_4N_2 ring joined to a subsidiary carbon ring, gives an empirical formula (with hydrogen attachments) of $C_8H_8N_2$, which can be broken into three H_2 molecules together with one or other of the pairs (HCN, HC_7N), (HC_3N, HC_5N). These are just the linear molecules given in Table 8.1. Pairs with odd numbers of C atoms are more stable than those with even numbers, and so would be favoured during a break-up process—favoured above (HC_2N, HC_6N), (HC_4N, HC_4N).

It seems possible that all the essential ingredients of life—polysaccharide chains, sugar-phosphate chains, amino acids, nucleotide bases, porphyrins, the carotenoids, were all added to the solar system by the apple-corer process of Chapter 8. The materials of life could thus have lain immediately to hand, waiting to be assembled like a model made out of the components of a Meccano set. And the origin occurred where? On the Earth one feels inclined to answer. But the comet-like carriers of the life-forming material to the Earth look themselves to be remarkably propitious sites for the origin of life itself. The time-sequence of events in the first tens of millions of years in the history of the solar system is especially relevant in this respect. Solid cometary-type nuclei of water ice were already present before the incidence of the apple-corer pick-up of interstellar material, and it would be on top of such nuclei that shells of organic material would be deposited. What we may next ask would be the physical and chemical evolution of such bodies?

The main physical evolution lay of course in a gradual accumulation into the planets Uranus and Neptune, but before Uranus and Neptune were fully formed some of the planetesimals would be sprayed into a giant halo surrounding the solar system. This would arise inevitably from close encounters between the planetesimal comet-type objects and the forming planets themselves. Bodies within such a giant halo would be subject to further perturbations from the nearest stars, which may well have been much closer at that time than the nearest stars are today—since the solar system was quite likely born in a cluster of stars. Perturbations from the nearest stars would sometimes cause the orbits of planetesimals to acquire perihelion distances $> \sim 100 \, \text{AU}$, and such planetesimals would then form more or less permanent attachments to the solar system, becoming the system that we recognize today as the reservoir of comets.

It is likely that chemical energy was available to be released by the skin of organics that settled on top of the icy nucleus of each cometary object. The glucose links in cellulose, for example, yield energy through

$$6H_2CO + 6H_2O \rightarrow 6CO_2 + 12H_2 + \text{Energy.}$$

Chemical energy released gradually near the boundary between the icy core and the organic skin could melt the ice. For this, the skin would need to be a few hundred metres in depth, to generate a pressure adequate to permit the existence of liquid water, but such a depth is quite consistent with the above estimate of the amount of organics that could have been added to the solar system by the apple-corer pick-up process. The further release of chemical energy, as for instance in glycosis, would heat the water, and with a mixing of water and organics the situation would correspond closely to Darwin's famous postulate of a 'warm little pond':

'... if (and oh what a big if) we could conceive in some warm little pond ... that a protein compound was formed ready to undergo still more complex changes ...'

The thermal conductivities of organics and of water are low enough for the cometary ponds, at depths of several hundred metres, to have remained warm for tens of millions of years, surely an excellent situation for the organics to have become organized into primitive life forms. Surely too, a terrestrial situation would be far less favourable. On the Earth, the oceans would have tended to swallow up the primaeval organics, reducing them to low concentrations, thereby making difficult any complex associations between one form of material and another—a difficulty which does not exist at all for the cometary situation, where the amounts of the organics and of water are comparable with each other.

It is still necessary to understand how life, arising in a cometary object, could emerge from it. Gravitational perturbations must also have produced orbits for a proportion of such objects with perihelion distances $< \sim 1$ AU. Even today, this happens for a few comets each year. In the first few hundred million years, before Uranus and Neptune were fully formed, planetesimals must have approached to within 1 AU of the Sun far more frequently than comets do nowadays. At each perihelion passage a little of the skin of a cometary-type object in such an orbit would be peeled away. Primitive life-forms originating below would thus be brought closer and closer to the surface. Surrounding water would become frozen into ice as the surface was approached. Then, with sufficient evaporation of the ice living cells could at last come free, perhaps to be quickly destroyed by sunlight, perhaps to be shielded from destruction by inert solid material, perhaps even to survive in the sunlight. If the Earth happened to be nearby, a surviving cell—or better many cells—could become entangled in the terrestrial atmosphere. Some would be slowed gently and would survive the slow fall down to the Earth's surface, there to find that a supply of the chemical foods to which they are accustomed had arrived already ahead of them. Survival until the chemical foods became exhausted was now a possibility, but not indefinite survival, unless a new trick could be learned.

Life is known to have passed from anaerobic (non-oxidizing) to aerobic (oxidizing) conditions, and there has been much controversy why, how, and when this transition took place. On the present view there is little alternative to fixing the transition at the moment of transition from cometary object to Earth. But with the transition completed continuing survival then depends on the trick of photosynthesis. The trick must be learned before the Earth's supply of chemical foods is used up, or before a continuing rain of foods from cometary objects onto the Earth falls off to an insubstantial trickle. Photosynthesis recovers the situation, by enabling the supply of foods to be regenerated *in situ*. With this crucial step, life may be said to have made the uneasy transition from its cometary home to the Earth.

The picture requires life to have arisen frequently inside the cometary bodies. If it happened inside only one body the chance of that particular body transferring living cells to the Earth would be very small; and even if cells were transferred the chance of those particular cells happening to solve the photosynthesis problem would probably also be small. But if many cometary objects each produced a profusion of living cells the situation is otherwise. Sooner or later one or other of them will shed cells onto the Earth, and if on the first occasion when successful shedding occurs the photosynthesis problem is not solved, further occasions will be coming along. If one is to have the present picture, this is the way it must surely be. The picture is strikingly similar to the scattering of seeds in the wind. Few are destined to take root, but so many are the seeds that some among them manage to do so.

Appendix
Abundances of
Nuclei

Element	A	% Abundance	Abundance
1 H	1	~100	3.18×10^{10}
	2		5.2×10^{5}
2 He	3		$\sim 3.7 \times 10^{5}$
	4	~100	2.21×10^{9}
3 Li	6	7.42	3.67
	7	92.58	45.8
4 Be	9	100	0.81
5 B	10	19.64	68.7
	11	80.36	281.3
6 C	12	98.89	1.17×10^{7}
	13	1.11	1.31×10^{5}
7 N	14	99.634	3.63×10^{6}
	15	0.366	1.33×10^{4}
8 O	16	99.759	2.14×10^{7}
	17	0.0374	8040
	18	0.2039	4.38×10^{4}
9 F	19	100	2450

Element	A	% Abundance	Abundance
10 Ne	20	(88.89)	3.06×10^6
	21	(0.27)	9290
	22	(10.84)	3.73×10^5
11 Na	23	100	6.0×10^4
12 Mg	24	78.70	8.35×10^5
	25	10.13	1.07×10^5
	26	11.17	1.19×10^5
13 Al	27	100	8.5×10^5
14 Si	28	92.21	9.22×10^5
	29	4.70	4.70×10^4
	30	3.09	3.09×10^4
15 P	31	100	9600
16 S	32	95.0	4.75×10^5
	33	0.760	3800
	34	4.22	2.11×10^4
	36	0.0136	68
17 Cl	35	75.529	4310
	37	24.471	1390
18 Ar	36	84.2	9.87×10^4
	38	15.8	1.85×10^4
	40		~20?
19 K*	39	93.10	3910
	40		5.76
	41	6.88	289
20 Ca	40	96.97	6.99×10^4
	42	0.64	461
	43	0.145	105
	44	2.06	1490
	46	0.0033	2.38
	48	0.185	133
21 Sc	45	100	35
22 Ti	46	7.93	220
	47	7.28	202
	48	73.94	2050

Element	A	% Abundance	Abundance
22 Ti (*continued*)	49	5.51	153
	50	5.34	148
23 V	50	0.24	0.63
	51	99.76	261
24 Cr	50	4.31	547
	52	83.7	1.06×10^4
	53	9.55	1210
	54	2.38	302
25 Mn	55	100	9300
26 Fe	54	5.82	4.83×10^4
	56	91.66	7.61×10^5
	57	2.19	1.82×10^4
	58	0.33	2740
27 Co	59	100	2210
28 Ni	58	67.88	3.26×10^4
	60	26.23	1.26×10^4
	61	1.19	571
	62	3.66	1760
	64	1.08	518
29 Cu	63	69.09	373
	65	30.91	167
30 Zn	64	48.89	608
	66	27.81	346
	67	4.11	51.1
	68	18.57	231
	70	0.62	7.71
31 Ga	69	60.4	29.0
	71	39.6	19.0
32 Ge	70	20.52	23.6
	72	27.43	31.5
	73	7.76	8.92
	74	36.54	42.0
	76	7.76	8.92
33 As	75	100	6.6

Element	A	% Abundance	Abundance
34 Se	74	0.87	0.58
	76	9.02	6.06
	77	7.58	5.09
	78	23.52	15.8
	80	49.82	33.5
	82	9.19	6.18
35 Br	79	50.537	6.82
	81	49.463	6.68
36 Kr	78	0.354	0.166
	80	2.27	1.06
	82	11.56	5.41
	83	11.55	5.41
	84	56.90	26.6
	86	17.37	8.13
37 Rb	85	72.15	4.16
	87		1.72
38 Sr	84	0.56	0.151
	86	9.86	2.65
	87		1.77
	88	82.56	22.2
39 Y	89	100	4.8
40 Zr	90	51.46	14.4
	91	11.23	3.14
	92	17.11	4.79
	94	17.40	4.87
	96	2.80	0.784
41 Nb	93	100	1.4
42 Mo	92	15.84	0.634
	94	9.04	0.362
	95	15.72	0.629
	96	16.53	0.661
	97	9.46	0.378
	98	23.78	0.951
	100	9.63	0.385
44 Ru	96	5.51	0.105
	98	1.87	0.0355
	99	12.72	0.242

Element	A	% Abundance	Abundance
44 Ru (*continued*)	100	12.62	0.240
	101	17.07	0.324
	102	31.61	0.601
	104	18.58	0.353
45 Rh	103	100	0.4
46 Pd	102	0.96	0.0125
	104	10.97	0.143
	105	22.23	0.289
	106	27.33	0.355
	108	26.71	0.347
	110	11.81	0.154
47 Ag	107	51.35	0.231
	109	48.65	0.219
48 Cd	106	1.215	0.0180
	108	0.875	0.0130
	110	12.39	0.124
	111	12.75	0.189
	112	24.07	0.356
	113	12.26	0.181
	114	28.86	0.427
	116	7.58	0.112
49 In	113	4.28	0.008
	115	95.72	0.181
50 Sn	112	0.96	0.0346
	114	0.66	0.0238
	115	0.35	0.0126
	116	14.30	0.515
	117	7.61	0.274
	118	24.03	0.865
	119	8.58	0.309
	120	32.85	1.18
	122	4.72	0.170
	124	5.94	0.214
51 Sb	121	57.25	0.181
	123	42.75	0.135
52 Te	120	0.089	0.0057
	122	2.46	0.158
	123	0.87	0.056

Element	A	% Abundance	Abundance
52 Te (*continued*)	124	4.61	0.296
	125	6.99	0.449
	126	18.71	1.20
	128	31.79	2.04
	130	34.48	2.21
53 I	127	100	1.09
54 Xe	124	0.126	0.00678
	126	0.115	0.00619
	128	2.17	0.117
	129	27.5	1.48
	130	4.26	0.229
	131	21.4	1.15
	132	26.0	1.40
	134	10.17	0.547
	136	8.39	0.451
55 Cs	133	100	0.387
56 Ba	130	0.101	0.00485
	132	0.097	0.00466
	134	2.42	0.116
	135	6.59	0.316
	136	7.81	0.375
	137	11.32	0.543
	138	71.66	3.44
57 La	138		0.00041
	139	99.911	0.445
58 Ce	136	0.193	0.00228
	138	0.250	0.00295
	140	88.48	1.04
	142	11.07	0.131
59 Pr	141	100	0.149
60 Nd	142	27.11	0.211
	143	12.17	0.0949
	144	23.85	0.186
	145	8.30	0.0647
	146	17.22	0.134
	148	5.73	0.0447
	150	5.62	0.0438

Element	A	% Abundance	Abundance
62 Sm	144	3.09	0.00698
	147		0.0349
	148	11.24	0.0254
	149	13.83	0.0313
	150	7.44	0.0168
	152	26.72	0.0604
	154	22.71	0.0513
63 Eu	151	47.82	0.0406
	153	52.18	0.0444
64 Gd	152	0.200	0.000594
	154	2.15	0.00639
	155	14.73	0.0437
	156	20.47	0.0608
	157	15.68	0.0466
	158	24.87	0.0739
	160	21.90	0.0650
65 Tb	159	100	0.055
66 Dy	156	0.0524	0.000189
	158	0.0902	0.000325
	160	2.294	0.00826
	161	18.88	0.0680
	162	25.53	0.0919
	163	24.97	0.08099
	164	28.18	0.101
67 Ho	165	100	0.079
68 Er	162	0.136	0.000306
	164	1.56	0.00351
	166	33.41	0.0752
	167	22.94	0.516
	168	27.07	0.0609
	170	14.88	0.0335
69 Tm	169	100	0.034
70 Yb	168	0.135	0.000292
	170	3.03	0.00654
	171	14.31	0.0309
	172	21.82	0.0471
	173	16.13	0.0348

Element	A	% Abundance	Abundance
70 Yb (*continued*)	174	31.84	0.0688
	176	12.73	0.0275
71 Lu	175	97.41	0.0351
	176		0.00108
72 Hf	174	0.18	0.00038
	176	5.20	0.0109
	177	18.50	0.0389
	178	27.14	0.0570
	179	13.75	0.0289
	180	35.24	0.0740
73 Ta	180	0.0123	0.00000258
	181	99.9877	0.0210
74 W	180	0.135	0.000216
	182	26.41	0.0422
	183	14.40	0.0230
	184	30.64	0.0490
	186	28.41	0.0454
75 Re	185	37.07	0.0185
	187		0.0341
76 Os	184	0.018	0.000135
	186	1.29	0.00968
	187		0.0088
	188	13.3	0.0998
	189	16.1	0.121
	190	26.4	0.198
	192	41.0	0.308
77 Ir	191	37.3	0.267
	193	62.7	0.450
78 Pt	190	0.0127	0.000178
	192	0.78	0.0109
	194	32.9	0.461
	195	33.8	0.473
	196	25.3	0.354
	198	7.21	0.101
79 Au	197	100	0.202
80 Hg	196	0.146	0.000584
	198	10.2	0.0408

Element	A	% Abundance	Abundance
80 Hg (*continued*)	199	16.84	0.0674
	200	23.13	0.0925
	201	13.22	0.0529
	202	29.80	0.119
	204	6.85	0.0274
81 Tl	203	29.50	0.0567
	205	70.50	0.135
82 Pb	204	1.97	0.0788
	206	18.83	0.753
	207	20.60	0.824
	208	58.55	2.34
83 Bi	209	100	0.143
90 Th[a]	232	100	0.058
92 U[a]	235		0.0063
	238		0.0199

This table is taken from A. G. W. Cameron, *Space Science Reviews,* **15** (1970), 121.

[a] Abundances at the origin of the solar system, 4.6×10^9 years ago.

Index

Index